T0407754

FISH, FISHING AND FISHERIES

EFFECTS AND EXPECTATIONS OF TILAPIA AS A RESOURCE

Fish, Fishing and Fisheries

Additional books in this series can be found on Nova's website under the Series tab.

Additional e-books in this series can be found on Nova's website under the e-book tab.

FISH, FISHING AND FISHERIES

EFFECTS AND EXPECTATIONS OF TILAPIA AS A RESOURCE

MARCO AGUSTIN LIÑAN CABELLO
EDITOR
AND
LAURA A. FLORES-RAMÍREZ
CO-EDITOR

New York

Library of Congress Cataloging-in-Publication Data

Effects and expectations of tilapia as a resource / editor, Marco Agustin Liqan Cabello (University of Colima, Mexico and co-editor, Laura A. Flores-Ramírez (University of Colima, Mexico)
pages cm. -- (Fish, fishing, and fisheries)
Includes bibliographical references and index.
ISBN 978-1-63463-307-9 (hardcover)
1. Tilapia. I. Liqan Cabello, Marco Agustmn.
SH167.T54E34 2014
639.3'774--dc23

2014039709

Published by Nova Science Publishers, Inc. † New York

Contents

PREFACE

In the past three decades, tilapia culture has grown dramatically around the world, carries the above benefits of economic and social order of great relevance in countries with tropical and subtropical environments worldwide. It is grown in approximately 85 countries around the world and a lot of the production obtained does not represent the endemic environment of this organism. Major industries are originated in the Far East, however, more and more it is cultivated in the Caribbean, Asia, Indonesia, many countries in America and, recently, in countries where systems are obtained by warm water by artificial means. Commonly, it is cultivated in ponds, cages and flooded rice fields.

Most tilapia species can grow in brackish water and some varieties have adapted to seawater, grows rapidly with feeds with lower protein food, tolerate higher levels of carbohydrates than other carnivorous farmed species also tolerate diets with higher percentage of vegetable protein, and is easy to play and grow in environments of low water quality and in systems with low to high level of investment and technology. All these features makes the tilapia a suitable specie for crop in most developing countries.

In many countries where the cultivation of tilapia has not been done from the perspective of system, would be best suited for growth planning and environmental management tool supported the activity of this species generally considered exotic. In many development countries culture originated from the identification and exploitation of diverse geographical / biological factors (the coastal area, water availability, introduction of exotic species, etc.) and not so much by the existence and implementation of a or directed integral policy program. Currently tilapia can be considered as a resource not only for aquatic environments but also brackish and marine, which has not been fully exploited. In many cases due to the lack of a management plan in countries where it is present. As a result of this, are the many problems it faces at the organizational level, water use, crop yield, such socioeconomic conflicts between the exercise of aquaculture and agriculture among others.

Effects and Expectations of Tilapia as a Resource has gathered internationally known experts who have prepared original communications in different thematic on the use and development of tilapia resource. Experiences related to culture, broodstock management, identifying plans and strategies technological reconversion works are presented; evidence of acclimatization in lake systems; further comprising a singular study of expectations and use of tilapia skin, among other chapters that will be of great interest to improve the opportunities for use of this singular resource.

In: Effects and Expectations of Tilapia … ISBN: 978-1-63463-307-9
Editors: M. A. Liñan Cabello et al.

Chapter 1

THE USE OF FISH-SKIN AS A PRODUCT FOR LEATHER MANUFACTURING IS AN UNDER-DEVELOPED BUSINESS OPPORTUNITY IN MEXICO

***José Martín Calvillo Mares*[*], *Elisa López Alanis*[†], *José Alfredo Rosas Barajas*[‡], *Víctor Ramírez González*[§] *and Marta Matsumoto Ramos*[**]**
CIATEC, Fracc. Industrial Delta, León, Gto. Mexico

ABSTRACT

"The use of fish-skin as a product for leather manufacturing is an under-developed business opportunity in Mexico"

Depending on the natural structure of the fish-skin, once transformed to leather, the product is aesthetically elegant. The natural elegance is further accentuated in fish-hides which have medium to small scales. The finished fish leather product when combined with knowledgeable craftsmen has significant artistic potential.

[*] jcalvillo@ciatec.mx.
[†] elopez@ciatec.mx.
[‡] arosas@ciatec.mx.
[§] vramirez@ciatec.mx.
[**] mmatsumoto@ciatec.mx.

Presently, this potential leather source is under-utilized in Mexico, due to an inadequate supply chain, in part due the complex manufacturing process required to transform fish-skin into leather, requiring a high-level of tannery aptitude, as well as, familiarity with the organic compositions of fish-skin.

In this paper, we will discuss the current market and future potential opportunities for leather fish-skin products in Mexico. In addition, we will describe the standard tanning process for fish-skins, while citing examples of value added products.

INTRODUCTION

Production and Trade of Fish Skin in Mexico

The use of fish skin in México is not fully quantified; currently there is little market data regarding the production and foreign trade.

With a market of more than 100 different species[1], in Mexico, during the Christian religious observance of "Lent" which occurs over an eight-week period, culinary consumption of fish is estimated at 250,000.00 tons, with a variation of 100 different species. The species of highest consumption in Mexico is the Sardine which approximately 700,000 tons[2] are captured annually. Other significant species include: Shark at 36,000 tons[3], Squid at 30,000 tons and Tuna at 150,000 tons annually (2013 statistics)

Presently in Mexico, there are no statistics which differentiate the different uses for individual species, as fish-skin production is estimated, as having presenting a significant opportunity for further exploitation.

A summary of the production of fish species is described in the following table:

The most implemented species in Mexico is the Carp with an of 250 grams per unit and an average size of 20-25 cm, of which the skin is considered a waste product, with the bulk of the residues converted into flour

---

[1] CONAPESCA; http://www.conapesca.sagarpa.gob.mx/wb/cona/18_de_marzo_de_ 2014_ mexico_df

[2] *Dirección de Estadística y Registros Pesqueros*, Sagarpa, http://www. conapesca.sagarpa. gob.mx/wb/cona/seguimiento_mensual_de_la_produccion_de_sardina_20

[3] FAO; *Servicio Nacional de Información de Mercados*

The Carp immediately consumed can be frozen for up to 7 days[4]; but eventually the target is food and not exploitation of the rest of the sub-products or residues.

Table 1. Capture 2012

SPECIES	VOLUME (KG)	VALUE (PESOS)
DogFish	4.507.848	76.735.019
Tuna	97.512.600	1.108.075.543
CatFish	4.382.095	115.089.219
Shark	16.766.322	304.336.214
Mojarra	74.126.299	1.384.183.511
Carp	26.177.188	325.044.900

Source: CONAPESCA, RNPA, SAGARPA, MEXICO 2014.

Skins Market in Mexico

The demand for fish skins has been steadily increasing. There is a growing acceptance of fish-leather for the production of articles or items such as vases, cup holders, centerpieces, wallets, belts, boots, etc., of which these products are manufactured in countries like Peru and Costa Rica, Colombia, Mexico and Ecuador[5].

The species most commonly used for fish-skin leather include: Tilapia, Shark, Skate, Trout, Salmon, among others.

Brands like New Balance footwear (tennis), Adidas and Puma among others, are betting on using fish skins in some applications.

There is ever greater presence of fish skin used in the manufacture of clothing as jackets, shoes, and clothes made by designers.

Tilapia Market in Mexico

On an international basis the fish species of Tilapia is the second most implemented in the production of aquaculture feed, over taking Salmon and Catfish.

[4] "La carpa y su manejo", SAGARPA, (colección Nacional de Manuales de Capacitación Pesquera), Delegación Tlaxcala, México. 1994.

[5] Moda y mar, "la piel de pescado" enero 2014. http://moda111.blogspot.mx/ search/label/leather.

Table 2. Imports of Fish Species in Mexico

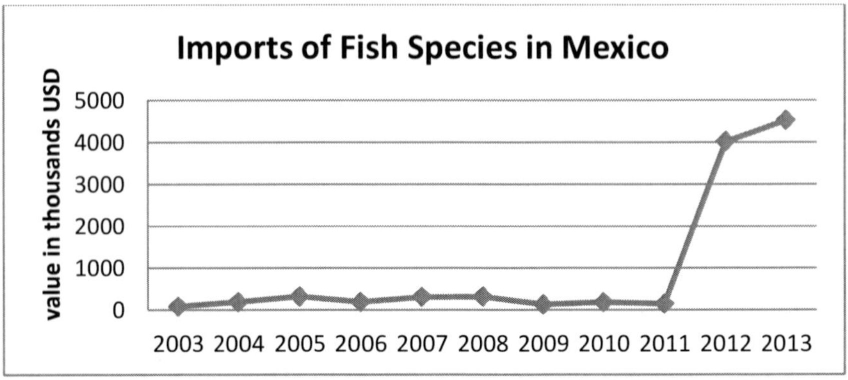

Imports of Fish Species in Mexico

value in thousands USD

5000
4000
3000
2000
1000
0

2003 2004 2005 2006 2007 2008 2009 2010 2011 2012 2013

Source: Secretaria de Economía, SIAVI 4; 2014.

Table 3. Export of Fish in Mexico

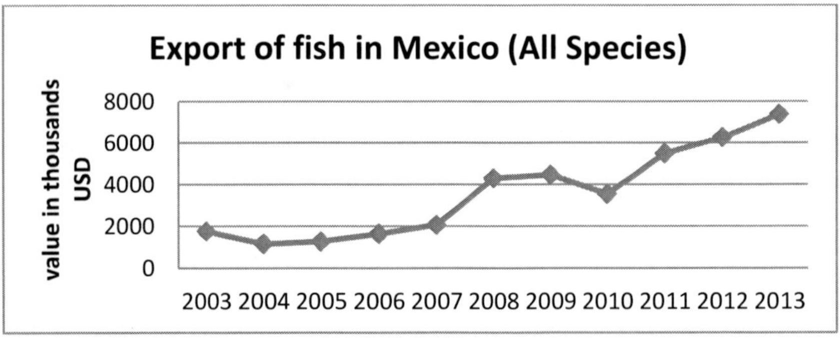

Export of fish in Mexico (All Species)

value in thousands USD

8000
6000
4000
2000
0

2003 2004 2005 2006 2007 2008 2009 2010 2011 2012 2013

Source: Secretaria de Economía, SIAVI 4; 2014.

During the World Congress of Tilapia in 2013, the estimated data cited that 4 billion metric tons are annually manufactured, with China, Indonesia and Egypt as the leading manufactures and Mexico ranking within the top 10 producers of Tilapia.

Approximately 10% of tilapia is produced in fish farms (aquaculture).

The largest importers of Tilapia in 2013 were the United States, Mexico, Saharan Africa, Russia and the European Union.

According to a 2010 study conducted by the “Committee System Tilapia Production of Mexico, A.C.” the domestic production of Tilapia ranked second with 71,018 tons manufactured.

One of the main purposes of the committee is to strengthen the supply chain by increasing productivity, which then will support the market growth objectives.

The study is a prospective analysis of 2020 on the tilapia market in México, highlighting some viability aspects, including the methodological process for integrating of a robust supply chain.

The analysis concerning the tilapia, for 2006, estimated that 30% of world fisheries were used for different purposes than food[6].

Internationally Mexico ranks as one of the top aquaculture manufacturers of shrimp, tilapia and oysters, with 25% of the market being Tilapia.

Almost 100% of domestic aquaculture production is intended for human consumption.

Aquaculture production in Mexico is almost 100% intended for use as food, the industrial production is not reported. In the case of tilapia in Mexico, is used 100% of the production for human consumption.

The domestic consumption of aquaculture products per capita increased by 28% between the years 2004 and 2008, with an average of 14 kg(s) per person.

At least 60 species of Tilapia have been documented; in 2009 international Tilapia production was estimated on 1.15 million tons with moderate growth increase.

Current market interests include the exportation of Tilapia to Latin American Countries, though in the case of Brazil, Mexico and Colombia, domestic production has satisfied market demand.

The largest Latin America producers of Tilapia are Brazil, Honduras and Colombia.

Ecuador is the leading producer of Tilapia for export to the United States with production ranging from 20,000 tons to 90,000 tons annually.

For Mexico, the Caribbean and the United States Tilapia farming is relatively new[7]. In Mexico Tilapia aquaculture was introduced during the 1960s. Tilapia production in Mexico has increased gradually; by 2003 a production of 61,500 tons was recorded, in 2008, 71 thousand tons[8], were produced.

Tilapia in Mexico is primarily from China, Taiwan and the United States, especially in the form of fish steaks for human consumption.

---

[6] FAO, 2009.

[7] Tilapia 2020 "Prospectiva del Sistema – Producto Nacional de Tilapia en México". 2010.

[8] (CONAPESCA, 2008).

More than 36,000 tons of Tilapia is imported from China, whose aquaculture performance is 5 to 8 tons per hectare, over a period of 8-10 months of harvest.

In Mexico significant demand for Tilapia which has coincided with a modest increase in domestic production. The current trade balance is negative for Mexico; by the volume and value of exceeding 140 million USD, therefore highlighting an opportunity for increasing domestic growth to satisfy domestic demand.

The domestic market for Tilapia mainly consists of fresh, unfrozen fillets, sold in the major urban centers of Mexico City and Guadalajara, via intermediaries who import the fish from cultivation sites located in Michoacán, Veracruz and Tabasco.

In consideration of the increased demand for manufactured leather goods containing Tilapia skin, which remains underexploited and due to the cataloging of the fish hide as a waste material, the market opportunities for fish leather remains vast and undefined.

ADDED VALUE, DESIGN AND INNOVATION WITH FISH SKIN

One of the primary objectives should be the development of a production and supply-chain process that transforms the raw waste material into a value added product. Use of the recovered fish skin further expands the productive aquaculture chain.

The finished leather fish-skin is stylistically exotic comparable to the skin of other reptiles, the main aesthetic being the vacant cavities which on contained the scales. The size and richness of the texture depends on the species and manufacturing process applied to the leather. Its main feature is the natural aesthetic design of the cavities left by the scales of the fish. The variety of sizes and richness of texture reliefs depends on the type of scales species and the finishing process applied on the leather.

The size of the skin depends on the species and weight at the time when the fish is harvested. A 750 grams Tilapia fish usually will provide a square decimeter per-side. To maximize the yield of the skin during harvesting, a proportional cut must be taken from the entire material, thereby minimizing waste. The modular pieces of fish-leather are joined together with glue or stitching depending on the artistic outcome desired.

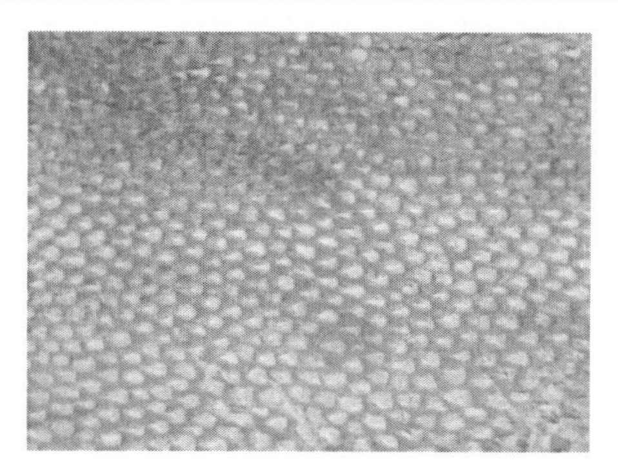

Figure 1. Dorado skin grain pattern.

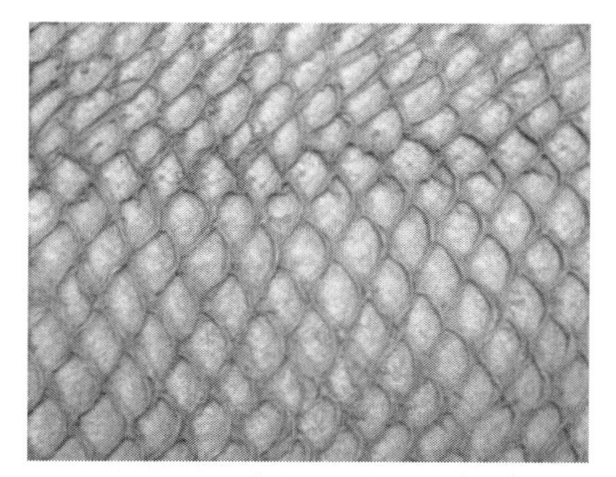

Figure 2. Tilapia skin grain pattern.

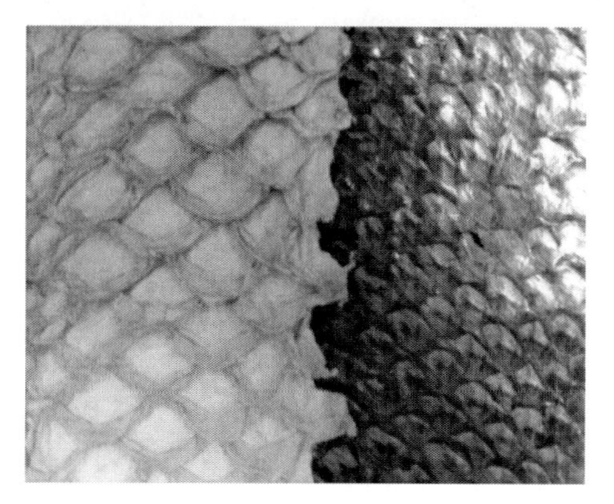

Figure 3. Snook Fish leather in crust and with glaze finish.

Figure 4. Carp Fish leather in crust and with glaze finish.

Figure 5. Shoulder bag made with Tilapia Leather. Design Elisa López; manufacturing development workshop, CIATEC, León, Guanajuato.

Figure 6. Cowboy boots made in patchwork, with Snook fish from the front to the heel. Made in Leon, Guanajuato, Mexico. By Rio Grande Boots.

Figure 7. Box of Tilapia Skin Panels with contrast stitching union.

Color is one of the most significant considerations leather design and with fish-skin there are no restrictions on colors which can be implemented, providing designers with ample design opportunity based on consumer preferences.

The color is seen differently depending on how the surface is textured, the more extensive the texture is, the richer and fuller the color will appear.

The strength of the fish skin is the result of the fibril entanglement and thickness. In physical testing, the tear strength, a 1.25 thick Snook skin has a resistance range from 12.6 to 7.1 kg and a 0.5 mm thick Tilapia skin has a resistance range from 3.2 to 2.1 kg., in both cases, the highest values occur longitudinally. For comparison, the Mexican Standard of Safety Shoes (NMX-S-051-1989) determines the materials used in the construction of shoes, a 10 Kg minimum and for the material used as lining 3.0 kg and fish leather is equally as tough as cowhide, pig or goat with equal thicknesses. However, it is recommended that in order increase the strength and shape of the skin, especially when worn in an adverse environment, the addition of an interlining is recommended.

Figure 8. Color Palette.

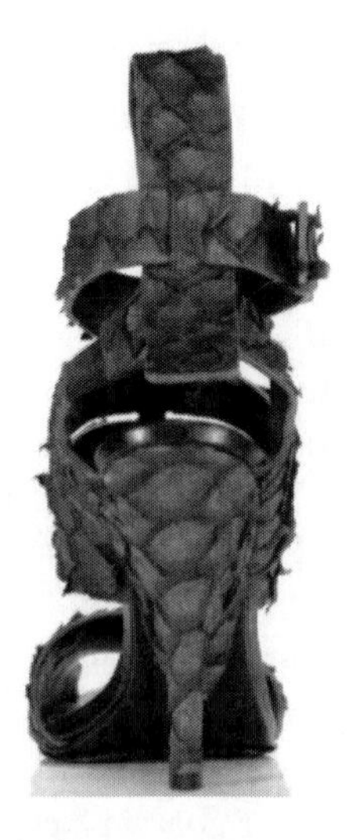

Figure 9. Slipper made of fish skin, by Alexander Wang, Spring 2012.

Presently, the major fashion houses focus their designs accessories. Due to increased media attention footwear companies now have greater exposure in the fashion industry, once only reserved for accessories.

Recently, fish leather has made appearances on the footwear and accessory collections of famous fashion designers.

Alexander Wang of New York was honored with the CFDA (Council of Fashion Designers of America, Inc.) as the Best Accessory Designer, which was subsequently followed by the launch of the Prêt-à-Porter collection where the fish leather had a leading role in the design of handbags, wallets and shoes.

But What Makes a Product Successful in the Market?

The implementation of fish leather into wearable fashion merchandise and accessories is increasing driven by incentive of the designers to produce modern and alluring compositions.

The implementation of fish leather into wearable fashion merchandise and accessories is increasing driven by incentive of the designers to produce modern and alluring compositions.

Basic Fish Skin Composition and Process

Fish have a natural epithelial skin layer which is derived of cellular constitution without the presence of blood vessels. The epithelial layer contains a mucous lining of glandular origin which may have greater or less abundance depending on the fish species. The mucous lining serves as a protective shield serving protective functions uniquely found in fish species.

The mucosa acts as a shield, by preventing the penetration of pathogens agents (bacteria, parasites, fungus). Moreover, the mucosa protects the skin from abrasion caused by adverse water conditions (pH modify, salinity, temperature, etc.).

When the water conditions exceed the species metabolic level, the mucous is released and the organism responds producing an increased quantity, with greater density or viscosity as required, over time the mucous naturally.

Other functions of the mucus lining include: the capacity to regulate water permeability, the maintenance of a proper osmotic pressure, in the form of a filter between two different concentration levels (a diluted state to a concentrated state). When the lining is compromised the fish can become dehydrated or over-hydrated then the surrounding water.

The fish skin has two primary layers the epidermis which is derived from the embryonic ectoderm and the dermis (or corium) is derived from the mesoderm and neural crest.

The epidermis is composed of several layers of flattened cells. The thickest layer is the active growth and multiplication zone (germinal layer functioning as a boundary between the body and the environment, so it has a very important role as it provides protection, fluid exchange and sensitivity. The epidermis serves a significant function by minimizing the water skin flux in conjunction with specialized cells in the gills.

The dermis is unique and is composed mainly of extracellular components.

Interspersed between the flattened cells of the epidermis are numerous mucous glandular openings which extend into the dermis. These glands produce the mucus that covers the body of the fish. The mucus decreases the frictional resistance while swimming, while fish expelling microorganisms and irritants when accumulated become toxic. In some species, the mucus clots and triggers the expulsion of solids. The mucous cells also function as a source of chemical communication.

The neural crest cells reside in the epidermis and dermis interface and in the proximal layers of the epidermis. These cells are called melanocytes and produce brown, gray or black melanin granules. Other tones can be produced by overlapping the melanin structures: iridescences refract light, chromatophores provide color and vascularization of the skin where the blood provides tones of pink or scarlet colors. The pigment migrations within the chromatophores are responsible for the color changes as fluctuating feature of the fish which is regulated by nervous and hormonal metabolic mechanisms.

Other feature of the fish skin includes glandular complexes are which less common, including: poison glands, light organs. Luminescence in fish is provided by the presence of bacteria in the skin of fish or by self-luminescence. Some extension areas of the skin such antenna with specialized functions, serve a sensory purpose typically implemented to locate and identify food. While other skin extensions have protective or ornamental functions. One of the most proximate features of the skin of the fish is the presence of scales, though it is important to note that all fish species have them. Depending on the species there intermediate states can be found in which the scales only partially cover the body. Other fish do not have scales but alternatively have bony features. Structurally, there are three different types of fish scales: cycloid, ctenoid, placoid, and ganoid. The scales are the organic construction material for many other biological features of fish such as: spines, plates, surface of the skull-bone and the operculum.

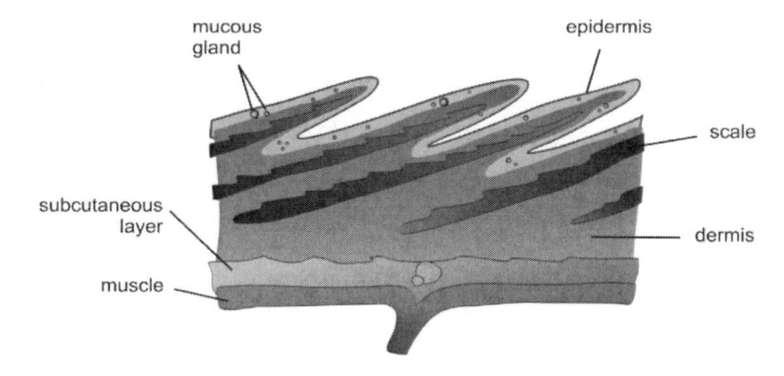

Figure 10. Transverse section of a fish skin.

Fish skin is comprised of the following layers:

a. Cuticle
b. Epidermis
c. Basal Membrane
d. Dermis
e. Hypodermis

a) Cuticle

It is the outermost layer consisting of mucopolysaccharides; this layer is extremely thin, only averaging a single micron thick. The physical consistency of the cuticle varies greatly from one species to species. The cuticle layer contains various defensive elements: specialized immunoglobulins, free fatty acids and a very important substance called lysozyme, all if which have anti-pathogenic properties. The cuticle is comprised of epithelial cells and mucous cells with cup-shaped shape.

b) Epidermis

The central component of the epidermis is the: “Malpighi” fiber cells. Unlike mammals, the epidermis is composed of cells capable of dividing by mitosis. The mucus excretory cells are found in the epidermis, with their density varying depending on the species and environment. In the epidermis are found other types granular cells. Furthermore in the epidermis layer appear cells such as lymphocytes and macrophages, both with immune defensive capacity. In the proximal region of the epidermis contains the germinative layer, which is a cell proliferation zone.

c) Basal Membrane

The Basal Membrane very thin layer and is basically a bonding layer located between the epidermis and the dermis, with no distinguishing feature.

d) Dermis

The dermis consists of two sub-layers: a spongy layer, which is formed by a sparse of collagen network, as this spongy layer contains pigment cells, leukocytes and scales. Below the dermis is the Compactum Stratum layer, which consists of a dense interwoven matrix of collagen, which is responsible for the strength of the skin. The pigment cells are called: "chromatophores" and are categorized by different types: (i) Melanophores are cells containing melanin having asteroid shaped features; (ii) Lipoforos contain soluble pigments, organic dissolvent´, of which there are two types: the Erythrophores (red pigments) and Xanthophores (yellow pigments), these two pigments cannot be synthesized by the fish, and must therefore be added to the diet; (iii) the Iridophores and (iv) the Leucophores, provide silver coloring, while Guanine, while the Leucophores which are located in the abdomen provide a whitish color.

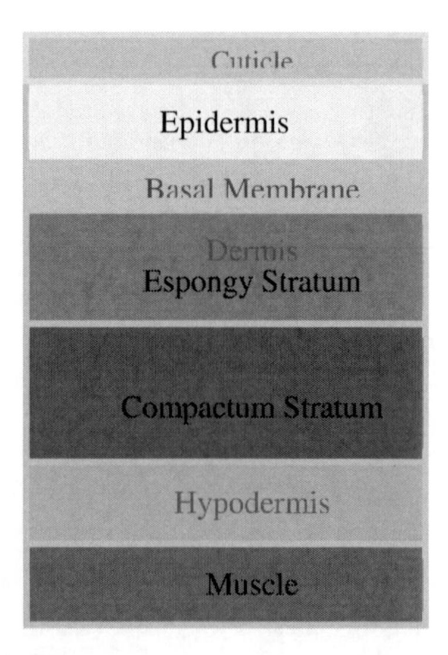

Figure 11. Layers of fish skin.

e) Hypodermis

The hypodermis layer is an enervated tissue, poorly vascularized and with adipose consistency. The hypodermis layer´s thickness is related to the species and the type of food consumed. This layer is often a place where infectious processes are started.

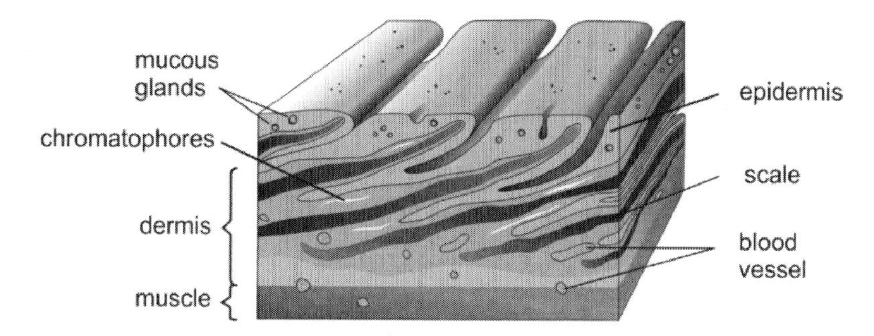

Figure 12. In this drawing we can see the sections of the fish skin and observe their characteristics.

Fish Skin

In general fish, dermis consists of a relatively thin upper layer of diffuse tissue area called compact layer. This area is rich in collagen fibers which are disposed in parallel form to the papillary layer and in a cross-linked form sheets, but not forming cross-linked networks as in the case of mammals.

Once deceased, the mucus is no longer offers effective protection and after a certain time, bacteria that consume the nitrogen of the mucus as a nutrient, leading to the degradation of the epidermis.

Figure 13. A big raw tilapia skin.

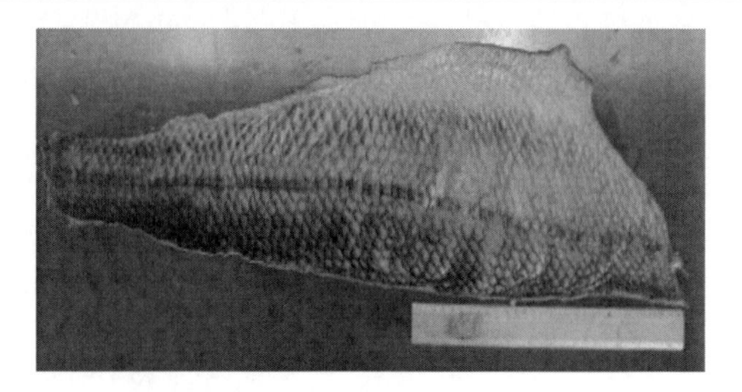

Figure 14. Raw snook skin.

The Collagen and Its Physicochemical

Collagen the main fibrous protein present in animals and are the primary connective tissue, constituting approximately one-third or more of the total body protein. The larger and heavier the animal is, greater the fraction of collagen is which contributes to the collective proteins in the body. It has been said that a cow retains its shape, mainly due to the collagen fibrils of the skin, tendons and other connective tissues. In cow skin, the collagen fibrils form a cross-linked network, with remaining portions in an almost perpendicular direction to the papillary layer.

The most abundant amino acid in fish skin is glycine. Glycine is collagen molecule formed by a chain consisting of 8 amino acids of glycine, plus 4 amino acids of proline, plus 2 amino acids hydroxyproline, plus 1 amino acid of arginine or lysine, plus 4 amino acid either tyrosine, aspartic acid, glutamic acid and histidine, and is structured repeatedly until it completes the polypeptide chain.

Although the collagens from different species vary in amino acid sequence, most contain about 35% glycine, 12% of proline and 9% of hydroxyproline, an amino acid that is rarely found in different collagen proteins. Proline and hydroxyproline differ from other amino acids because in its "R" group is a substituent in the amino group. The secondary structure is deduced to form a triple polypeptide chain which turns left, and is adhered by hydrogen bonds.

Figure 15. Side flesh of a tanned fish where the fibers are observed in a plane parallel to the surface and crossed almost 90 ° layers.

Figure 16. Fleshing machine.

The hydrogen bonds are formed from a carbonyl group consisting of a polypeptide chain and an amino group from an adjacent chain. The hydrogen bonds are common in the configuration of proteins and are the fundamental chemical basis for understanding the complex behavior of collagens reaction to pH, temperature and other physicochemical variables.

Furthermore, polypeptide chains of collagen contain a the hydroxyproline amino acid, permitting the formation of alternative of hydrogen bonds via a junction, such as a carbonyl group located in the pyrrole ring of hydroxyproline, providing further stability to the secondary structure by comparison to other proteins. It should be noted that with an increased number of hydrogen bonds, an increased temperature is required for complete denaturation.

Tanning Skin Fish

Fish skin has a smooth layer with moderate pigmentation, to which the scales are firmly adhered.

In comparison to bovine and swine hides, harvested fish-skins are very small, and, therefore, any fleshing process should concentrate minimizing waste.

The skin should be classified by species, size and pigmentation.

The desirable characteristics of the fleshed fish-hides prior to tanning should include:

1. The hides are clean and healthy
2. Flesh content is minimized
3. Irregularities due to poor fleshing techniques are minimized
4. The hides should be uniform in size

The tanning process occurs on 6 stages, which only the first stage occurs in the slaughter house, and the subsequent 5 stages occur within the tannery facility, these stages are included the: Beam House, Tanyard, Retan, Color, Fatliquor, and Finishing. Each stage is divided in 17 processes that are listed in Figure 17.

Figure 17. Stages of tanning process.

Conservation

a) Skinning and cleaning
b) Fleshing
c) Conservation (salting, freezing)

Skinning

Risk of contamination should be minimalized during filleting and transport to the conservation section.

It is recommended once the skin is removed from the animal; the hide is immediately placed into clean and suitable containers to prevent contamination with the residual flesh that found on the ground surface of the slaughter house.

Fleshing

Fleshing is critical step in ensuring proper preservation. The first step of the fleshing is to extend and lie flat the skins on a clean table with the flesh-side facing upward. This is then followed by removing the excess flesh with a knife or scraper, using care to prevent puncture holes and terminate marks on the hide, thereby decreasing the value. Tails, thorns and thick regions which the salt cannot penetrate should be removed and can adversely affect the preservation process.

Conservation of the Hide

Types of Conservation

1. Brine

Brining consists of soaking the skins in a container of salt saturated water for later use, or until the salt has been absorbed, reaching equilibrium, followed by draining and storing cool climate controlled environment

2. Dry Salt

After filleting and fleshing, the skins are washed in water, and allowed to drain for 10 minutes. After draining, the hide is placed flesh-side upward and coated in fine grain salt. The quantity of salt used is equal approximately 50% of the total weight of the skin, and distributed uniformly. Afterwards the hide is further drained on an inclined table for a minimum of two hours. Lastly, a smaller quantity of salt is applied and the hides are stowed in pairs with the flesh sides facing inward.

It is advisable to apply 50% or more by weight of salt on the skin. After that, let the skin drain in an inclinated table for 2 hours at least.

3. Freezing

The washed skins are drained and are stowed in pairs with the flesh sides facing inward. Then stored frozen at a temperature below 0^0C.

4. Storage

Dry salted preserved hides should be stored in a controlled cool and dry environment and, if possible, refrigerated until use. Warm and fluctuating temperatures facilitate degradation. Special attention should be paid that no putrefaction has occurred. Poorly fleshed hides cause improper chemical saturation in the subsequent processes.

Beam House Operations

a) Soaking
b) De-hairing and Liming
c) Deliming, Bating, Bleashing, Defating.

Soaking

All wet processes must be carried out in wooden, plastic or stainless steel drums, at the correct speed and with different equipment for each stage of the procedure. Alternatively, if specialized drums are unavailable, plastic buckets with sufficient stirring may be implemented. The purpose of the soaking is to clean the hide while removing salt and other impurities.

Abundant water quantities should be provided to ensure sufficient saturation of the hide. When implemented in conjunction with a surfactant and a bactericidal will further accelerate the process by removing the natural fat and other contaminates. In addition, a small quantity of sodium carbonate or salt can be added to assist with dissolving the globular proteins. The surfactant is a helper for soaking, accelerates the process and partially removes the natural fat that, in joint with bactericidal product, leaves the skins clean of dirt. The purpose of the rehydrating the hides is to reestablish flexibility. Should the hides still have scales, a rehydration process is necessary followed by a dry rolling. Depending load size typical dry rolling times are between 30 minutes and 2 hours. Care should be taken during the drying rolling process to ensure that the hides do not get heated or damaged.

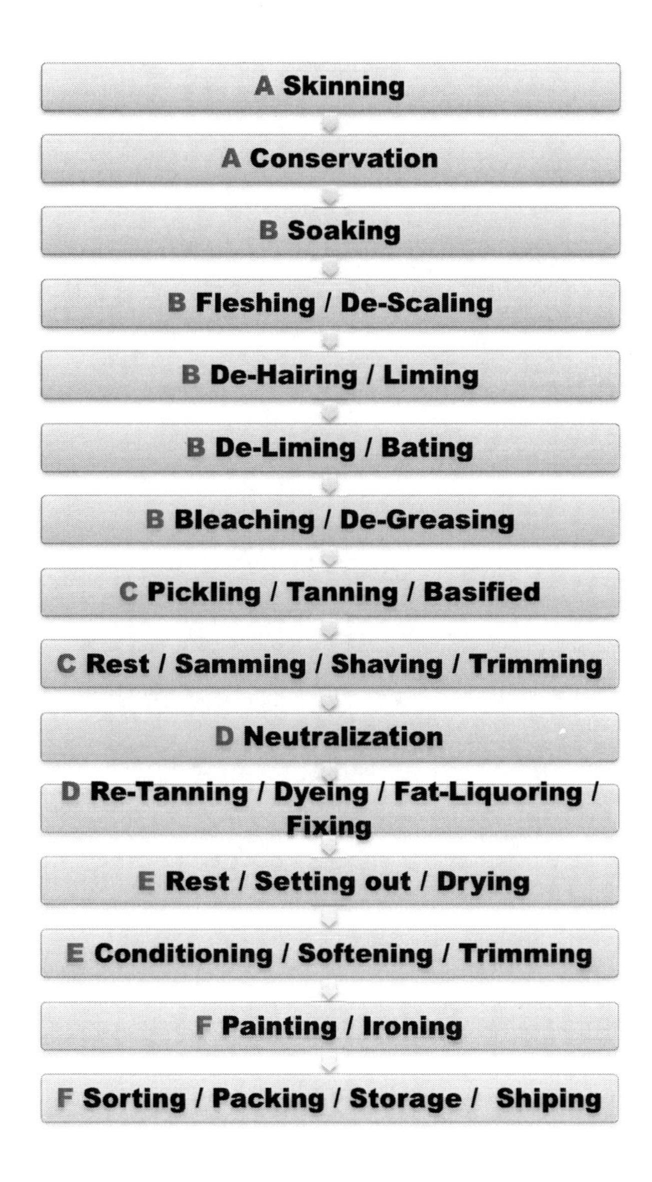

Figure 18. Diagram of tanning process.

Figure 19. Tanning pilot plant.

Figure 20. Tanning drum, made with stainless steel and clear acrylic face.

De-hairing (De-scaling) and Liming

The De-hairing process aims to remove the gelatinous pigmented layer along with any remaining scales. Combining sodium sulfide with a strong alkali (lime) triggers swelling, causing a charge repulsion between the protein molecules, allowing water to permeate.

Typically this process takes between 12 and 24 hours. To open the spacing between the fibers and achieve a uniform smoothness, depending on the species, the liming process can take place over a one to two day period. A proper liming process is essential for the remaining tanning and lubrication process

Figure 21. Pilot size wood tanning drum.

De-liming, Bating, Bleaching and De-greasing

The purpose of the de-liming process is to eliminate the lime absorbed by the hide, so that bating can take place. A swelling annulment is achieved when equilibrium between the liming bath and interior of the hide has been reached. In this process, sulfate salts or ammonium chloride, supplemented with some bisulfite are implemented. Phenolphthalein is used for process control, as a pH indicator which turns violet when a level higher than 8.5 is reached. The degreasing (rendering) action is achieved by the action of pancreating proteases and bacterial-type enzymes. After rendering the hides are washed once or twice with sodium bisulfite or for a stronger effect potassium permanganate, or other surfactants. This process usually takes over 1 to 2 hours using one of more baths, depending on species.

Tanning

a) Pickling
b) Tanning
c) Basified

Figure 22. Laboratory (Reactor) stainless steel tanning drum with opened door.

Pickling

The pickling process raises the pH to between 2.5 and 3.0 allowing the tan chroming to occur. It is important to note that the hides are sensitive to acids, so the implemented acids should be 10 times diluted in a saline medium, to prevent acid swelling and damage.

The addition of acid should be done while the drum is rotating and after the salt bath has achieved a density greater than 8^{a}Be. The drum should be rotating a low speed (10rpm) so that the hides receive a proper Consider explaining what "tapping" means achieving the wanted pH values, otherwise it may only achieve a: "dead tanning", whereby tanning has only occurred on the surface of the skin, failing to penetrate the interior fibers.

Tanning

The tanning of hides prevents putrefaction, which is anaerobic decomposition. Tanning can be achieved using different compounds including: vegetable tannins, mineral salts, such as chromium, aluminum or zirconium and synthetic tanning agents such as phenol derivatives, naphthalene or modified by aldehydes. These reagents have tanning action either as filler for fibrillar skin structure or as direct reaction on collagen. The tanning method implemented is dependent on the desired leather characteristics For example, a chrome tanning process will create leather tear-resistant, voltage resistant and temperature resistant properties, while the other types of tanning have decreased resistant. Other tanning components provide excellent properties such as: fullness, sweat-resistance, compactness, the permitting of exotic colorings (dyeing or whiteness). However, internationally more than 80% of hides tanned are processed with chromium, due to its versatility and low-cost.

Basified

The tanning process should be complemented with a locking tanning material. In the case of tanning with chromium, a weak alkali combined with organic tannins and a weak acid such as formic acid. Both materials must be diluted 1 to 10 in water. In both cases, the resulting pH should be around 4.0.

Wet Finishing

a) Shaving and Trimming
b) Neutralization
c) Rechroming
d) Dyeing
e) Fat Liquoring

Shaving and Trimming

After tanning, the skins must be stowed, separately, for one or two days to reach peak adhesion and completion of the chromium molecule. The purpose of this process is to give the leather the required final thickness. Lastly, the leather is processed in a downgrade machine, which is 50 inches width, after the leather has been mechanically drained by pressure between two rollers wrapped in felt. If a samming machine is unavailable to drain the leather, it is recommended to place the skins outdoors, in a shaded area for a few minutes until they reach 50% moisture or in absence of a measurement tool, or until the leather has a consistency form permits machining. Shaving removes the remaining residual flesh,. Should the proper equipment be unavailable for this process, it possible to adjust thickness manually using moisture resistant sandpaper or with emery.

Neutralization

While curing, the chromium adheres to the collagen fibers causing a release of acid, decreasing the pH of the leather, so a neutralization process is necessary for the interior surface.

The complexity of the tanning process is largely dependent on the type of leather being manufactured. An exceeded neutralization of the pH will loosen the fibers, resulting, poor penetration of the re-tanning, anilines and oiling, leaving the leather with problems like stains, brittleness and with drying and finishing issues. This process is completed using smooth alkalis such as sodium or calcium formate and sodium bicarbonate or ammonium. At the conclusion it is necessary to soak the leather in a bath, for 20 to 30 minutes, at this point, the final pH should be between 4.5 and 5.0 for firm leather and greater than 6.0 to smooth leather. The pH control is monitored using a potentiometer in the bath.

Re-Tanning

In the re-tanning process the fullness of the leather is increased, manipulating this process can result soft or hard, elastic or rigid and smooth or rough leather. This is achieved by adding specific materials like vegetable tanning agents, minerals, organics, resins or polymers, which do not override the character of chrome tanning process. There is a great diversity of products that serve as tanning agents which are mostly of a synthetic origin.

Dyeing

While curing, the chromium adheres to the collagen fibers causing a release of acid, decreasing the pH of the leather, so a neutralization process is necessary for the interior surface. Anionic dyes are the most commonly used and are suitable for dyeing re-tanned leather. The composition of the dye is adjusted according to the depth of penetration required, based on the thickness of the leather. Some dyes are blended with metals to increase their hardness, while others blended with acids to intensify the color.

Fat-liquoring and Fixing

The fat-liquoring process can be aligned to the dyed leather or achieved separately in a new bath, after the anilines are fixed. The objective is to give the leather softness desired; to achieve a uniform lubrication a variety of oils should be used to achieve a symmetrical balance thereby providing a soft, delicate touch with a natural appearance. Fat-liquors are typically anion-active, suitable for the manufacturing of water-oil emulsions, but also for cationic surface effects. To prepare the fat liquoring bath, the emulsified oil must be at least 5 times its weight in water, at a temperature of 60^0C-70^0C. If the emulsion is prepared in the drug with cold water in oil and diluted, the emulsion will break causing the grease to be deposited on the surface, resulting in greasy leather. A wide variety of oils may be implemented, animal oils provide high softness (especially oil fish and neats-foot), Vegetable oils provides a dry, medium-firm texture, while mineral oils facilitate a more solid leather. Sulfitation administers a deeper oil penetration and stability even in a salty or acidic media. The sulfated and sulfonated oils have less stable emulsions but provide greater fullness and surface effect, while synthetic oils and phosphates provide increased stability and robustness. At the conclusion of the process, a greasing is required to insure an adequate adhesion of the materials, typically using small percentage of formic acid or another equivalent active material. Allowing coating to penetrate the leather 15 or 30 minutes and then afterwards washing out the access for 5 or 10 minutes to remove any unbound material, doing so prevents staining.

Drying (And Pre-finishing)

a) Drying
b) Softening

Drying

At the conclusion of wet finishing process, and prior to commencing the drying process the leather is allowed to cure for 12 hours to ensure the complete reaction of the materials. Drying process significant modifies the characteristics of the leather as dictated by the reduction in humidity and surface contraction. Less obvious changes, include: variation in the isoelectric points, changes in the formation of the inter-fiber bonds, the migration of soluble substances which were not properly adhered to the surface. Should the leather be dried outdoors, the leather will shrinks, shrivel and becomes rigid.

To insure the desired features remain intact drying the drying process the leather should be placed on a flat, clean, uniform surface. Fast drying causes a low-quality leather, while a slow, controlled drying, results in high-quality leather. Prior to commencing the drying process, best manufacturing processes recommend draining and stretching the leather, using roller. Doing so can eliminate 50% of the permeated water. In addition, measures should be taken to prevent excessive shrinking of the leather, for example, the drying temperature should never exceed 40^0C. The completion of entire drying process should take no longer than 12 hours. The objective is to have an approximate moisture content of 12% to 14%. The temperature to work with the leather should never exceed 40^0C to prevent leather from drying out and get hard during the entire process, which should not be more than 12 hours.

Softening

Once dried, the next process is the softening. This process adjusts the smoothness according to the desired traits. Doing so, may require an additional step where the leather is moisturized and covered for two hours before applying friction or tension, by mechanical means or friction. The moisture content of the leather, prior to the softening process, should be uniform and around 14% and 18%. Also is possible to soften the leather implementing a dry tumbling method, of which the duration will depend on the desired softness a typical tumbling drying times lasts between 2 to 8 hours. Most of the time, this method of softening should conclude with a tensioning step, lasting approximately 2 hours, against a flat surface to decrease skin elasticity while smoothing the surface.

Finishing

The softened fibers are ready for the finishing process which consists of the application of an aqueous and solvent borne resin applied by brush or sprayed, in a direction which maximizes uniform absorption. Normally, the leather received two uniform light applications with a drying process in between the two applications base. Once dried, the leather is ironed at a low temperature with minimal pressure, on a heated glass plate at 60^0C and 60 atm. After ironing, the leather receives a third layer of light base, once dried, the leather is further protected by applying a layer of an aqueous top or glossy lacquer solvent, if necessary the process is concluded with additional ironing. Fish leather is has a high intrinsic beauty, so the finishing process should be the minimal necessary to achieve acceptable protection without risking deterioration of its natural beauty. The most natural finishing process are based on casein and waxes, and only reinforced by small quantities of special resins (dry touch), without pigments or with small amounts of organic pigments.

Below are formulas to tanning the skins used in the products presented. Although it described in a generic formula, we have noted example materials, without implying that they are the only, or have better properties against similar products. The formulas cited have been used for a long time, however, preliminary testing is highly recommended to make the necessary adjustments to the raw materials, prior to producing the initial first batches.

Recipes

ARTICLE	**Fish Skin**		**PROCESS:**		**1 Beamhouse**	
PROCESS	%	MATERIALS	Time Min.	Temp. °C	pH	Notes
Raw Material: Salted skins						
Pre-soaking	200	Water		t. a.		
	0.1	Bactericide				Busan 85 (Buckman)
	0.5	Surfactant				Borron JU (TFL)
Rest			60			
Move			60			
Drain						
Soaking	200	Water		t. a.		
	0.1	Bactericide				Busan 85 (Buckman)
	0.5	Surfactant	60			Borron JU (TFL)
Rest			30			
Move			60			
Rest overnight						
Control process						
Scaling and Fleshing						
Liming	200	Water		t. a.		
	1	Sodium Sulfide	45			
Rest			30			
	1.5	Sodium Sulfide				
	3	Calcium hydroxide	45			
Rest			30			
	3	Calcium hydroxide	45			
Rest			30			
Move			45			
Rest overnight						
Drain and Rinse						
Rinse	200	Water				
	0.5	Ammonium Sulfate	15			
Drain						
Deliming	200	Water				
	1.5	Ammonium Sulfate				
	1	Surfactant				Decaltal R (BASF)
	0.5	Sodium bisulfite				
	0.5	Surfactant	60			Borron JU (TFL)
Control Process						
Drain						
Degreasing	200	Water				
	0.5	Surfactant	30			Borron JU (TFL)
Drain						

Recipes (Continued)

ARTICLE	Fish Skin		PROCESS:		1 Beamhouse	
PROCESS	%	MATERIALS	Time Min.	Temp. °C	pH	Notes
Bleaching	200	Water				
	4	Salt				
	0.6	Potassium permanganate	10			
	100	Water				
	4	Salt	20			
Drain						
	200	Water				
	2	Sodium bisulfite	20			
	2	Sodium bisulfite				
	2	Salt	20			
Drain	200	Water				
	1	Sodium bisulfite	20			
Drain and Rinse						
Pickling	200	Water				
	15	Salt	15			
	0.5	Formic Acid	20			Diluted 1:10
	0.7	Sulfuric Acid	60		2.5	Add three parts every 15 minutes
Control process						Court
Rest overnight						
Tanning	6	Chrome Tann agent 33%B	60			Baycrom 33 (Lanxess)
Cutting Control						
Basification	0.45	Magnesium Oxide	8 hours		3.8-4.0	Cromeno FN-1 (TFL)
Control Tanning						
Drain and Stowed						
Rinse	200	Water		t. a.		
	0.3	Tensoactivo				Borron JU (TFL)
	0.2	Oxalic Acid	30			
Drenar						
Pretanning	150	Water		t. a.		
	2	Synthetic Chrome	40			Tanesco H (TFL)
Neutralized	1.5	Sodium formate		t. a.		
	1	Neutralizing agent	30			Tanigan Pakn (Lanxess)
	0.5	Sodium Bicarbonate	30			Control pH
Drain and Rinse						
Retanning	100	Agua		t.a.		
	2	Acrylic Retannage	30			Leukotan 1028
	6	Filler Retannage				Sellatan LV (TFL)
	6	Bleach Retannage				Sellatan FL Liq. (TFL)
Rinse	200	Water		t. a.		
	0.3	Tensoactivo				Borron JU (TFL)
	0.2	Oxalic Acid	30			

ARTICLE		**Fish Skin**	**PROCESS:**		**3 Wet Finishing**	
PROCESS	%	MATERIALS	Time Min.	Temp. ° C	pH	Notes
Drenar						
Pretanning	150	Water		t. a.		
	2	Synthetic Chrome	40			Tanesco H (TFL)
Neutralized	1.5	Sodium formate		t. a.		
	1	Neutralizing agent	30			Tanigan Pakn (Lanxess)
	0.5	Sodium Bicarbonate	30			Control pH
Drain and Rinse						
Retanning	100	Agua		t.a.		
	2	Acrylic Retannage	30			Leukotan 1028
	6	Filler Retannage				Sellatan LV (TFL)
	6	Bleach Retannage				Sellatan FL Liq. (TFL)
	2	Rosin Retannage				Tergotan RD (Trumpler)
	2	Proteinaceous Retannage	90			Trupotan TFP (Trumpler)
Fatliquoring	5	Mixture Fatliquor				Leder soft SGA(ILCD)
	2	Sulfited Fatliquor	45	50		Leder Soft MK (ILCD)
Fixing	0.8	Formic acid	20			Diluted 1:5
Drain						
Rinse	200	Water	5			
Drain						
Dyeing	200	Water				
	2	Dispersant Retannage	20			Sellasol TD (TFL)
	3	Dye	45			
	1	Formic acid	30			Diluted 1:5
Drain and Rinse						
	300	Water				
Fixing	0.8	Fixing Agent	20			Sellafix WS (TFL)
Drain						
Drying Fixed, Staking						

ARTICLE	**Fish Skin**	**PROCESS:**	**4 Finishing**
MATERIALS	Preparation		PROCESS
	1st	2nd	
Wather	55	55	Apply 2 x Preparation 1, Dry
UR 1701 (Polyurethane Resin Dispersion (Stahl))	25	20	Iron 60°C, 120atm
FI 1950 (Wax Dispertion (Stahl))	5	5	Apply 2 x Preparation 2, Dry
LF-4247 (Proteinaceus Binder (Central Kimica))	15	20	Packing

References

CUERONET. Luis A. Prado Pasos. Técnico Curtidor Especialista en pieles exóticas. Costa Rica

Functions of Fish Skin: Flexural Stiffness and Steady Swimming Oflongnose Gar *Lepisosteus Osseus* John H. Long, Jr, Melina E. Hale, Matt J. McHenry and Mark W. Westneat.

La piel de los Peces. Roberto Petracini.

zvert.fcien.edu.uy/nuevos_cursos/practico_03_peces_oseos.pdf.

In: Effects and Expectations of Tilapia … ISBN: 978-1-63463-307-9
Editors: M. A. Liñan Cabello et al.

Chapter 2

TILAPIA BREEDING PROGRAMS FOR FILLET TRAIT TO MEET MARKET AND CONSUMER PREFERENCES

Mohamed E. Megahed*

National Institute of Oceanography and Fisheries (NIOF),
Gulfs of Suez & Aqaba's Branch, Attaka, Suez, Egypt

ABSTRACT

This chapter presents the breeding plan for selective breeding program for Nile tilapia (*Oreochromis niloticus*). This is meant to be used as operational document describing the various work operations in a breeding program where families is produced in batches (Three cohorts per year) or alternatively once a year, with genetic links between the batches. To do so a locations where families can be produced all year around is an absolute condition. This breeding program should be dynamic, i.e., be regularly evaluated and updated in order to reflect experiences and results obtained when developing the breeding program. The overall objective is to assure an optimal use of all resources allocated to the breeding program for Nile tilapia to develop genetically superior Nile tilapia. One of the program gaols is to select for fillet color and fillet weight in Nile tilapia for consumer preferences, the program is currently evaluating the benefits from sloding tilapia as fillet and also the value of different fillet colors in terms of consumer preferences.

* aquageimprove@gmail.com.

INTRODUCTION

There is an increasing demand of fish in the world due to an increasing population, better economy and more awareness of healthy food. Since the capture fisheries have stagnated, fish farming has become the fastest growing food production system. While producing inexpensive animal protein, aquaculture may play an important role in alleviating pressure on natural fisheries that are overexploited or, in the best scenario, exploited at a maximum. Intensification of aquaculture in Africa can lead to further reduction of the nutritional gap and minimization of production cost while conserving fresh water, which is becoming a competitive resource in many parts of the continent. Aquaculture of tilapia has gained publicity in many parts of the world including Europe and the United States. To a great extent, this is due to the ease of their reproduction and farming, their wide range of environmental tolerances, and their market acceptability. In Africa, the above factors along with their natural abundance in most of the continent creates high hope that they will contribute to enhancing the role of aquaculture in fighting famine and malnutrition in many parts of the continent. Tilapia contribution to poverty alleviation can be enhanced significantly through increasing their use and production in aquaculture. The growing interest in aquaculture of tilapia around the world has led to the adoption of several breeding programs for improving its productivity [1, 2, 3, 4, 5]. Estimation of genetic parameters such as heritability and genetic correlations between important traits are immensely valuable in predicting the chances of success in improving performance, avoiding unnecessary problems with deterioration of other traits, and estimating the breeding values of individuals and predicting the response to selection. Upon designing a breeding program, both the characteristics and the needs of the existing aquaculture system must be seriously considered while setting the breeding goals and objectives. The current breeding program discussed in this chapter was designed to suit the purposes of Egyptian/African aquaculture where the main emphasis is on increasing production to alleviate poverty and fight the increasing famine in African communities.

As a result it is believed that substantial improvements in aquaculture production in Africa can be achieved through similar selection for faster growing fish in this region. Direct transfer of the improved strains from Asia has not been undertaken because of concern over the potential adverse impact on native germplasm in Africa and unknown effects of gene-environment interactions. The approach being adopted by the current breeding program has therefore been to adapt the technology demonstrated in Asia for use in the genetic improvement of tilapia in Africa, test other approaches, and study the potential environmental consequences of genetically modified stocks. There is a big, however growing need for intensified, well-structured work on fish genetics in Africa. The area is a relatively new one and big leaps can be accomplished therein and achievements are normally of a lasting nature for little or no additional cost. Model work that was developed in other parts of the world can be used to attain high potentials of success in similar African countries. Moreover, research topics that cover aspects of uniqueness in the African fish and/or environments can also be developed.

Here I may have a number of outputs that would benefit the target region in a number of ways:

a) Improved end products, e.g., higher performing lines, can be utilized at different levels depending upon the product and the technology used for its development. A faster growing line of Nile tilapia that will be developed from Egyptian stocks is useful only for aquaculture in Egypt. However, if the improved line is completely sterile or if sterility can be introduced into it, then the range of beneficiaries can be extended to include the entire target region.
b) Improvement technologies are usually transferable without major modifications for use on the same species and, in many cases, for use on several other species as well. This, however, requires that the necessary expertise exists in the recipient location.
c) Identification of possibilities for inclusion of additional improvement techniques through utilization of new technologies and/or developing better ways to assess and quantify the ecological impact of the new technology. Based on such evaluations, an informed decision can be made on the release possibilities of a given product for commercial use either regionally or sub regionally.

BIOLOGY OF TILAPIA

"Tilapia" is the common name applied to three genera of fish in the family Cichlidae: *Orecohromis*, *Sarotherodon*, and Tilapia. These genera are distinguished with different reproductive behavior: all Tilapia species are nest builders, where *Oreochromis* and *Sarotherodon* species are mouth breeders. In *Oreochromis* species only female practice mouth brooding, while in *Sarotherodon* both sexes are mouth brooders.

Figure 1. Nile tilapia (*Oreochromis niloticus*).

The species that are most important for aquaculture are all in the genus Oreochromis, including Nile tilapia (*O.niloticus*), Blue tilapia (*O.aureus*), Mozambique tilapia (*O. mossambicus*), and Zanzibar tilapia (*O. urolepis hornorum*). Nile tilapia (Figure 1) is the main cultured species in the world and accounts for about 90% of global aquaculture production of tilapia. Males of pure strains of Nile tilapia and hybrids with Nile tilapia as a parent are considered the fastest growing tilapia varieties. Nile tilapia performs well in tropical and sub-tropical areas (preferred water temperature between 2°C – 30 °C) and can be farmed in a range of production systems (i.e., extensive to highly intensive; mono and polyculture). Nile tilapia can grow to a maximum body size of 4,300 grams (60 cm body length).

The food intake of Nile tilapia will be significantly reduced when the water temperature falls below 20°C, and water temperatures below 11 °C are lethal. Optimum water temperature for reproduction is 25 - 29 °C, at which Nile tilapia becomes sexually mature after 5-6 months (> 10 cm). Nile tilapia is the least salinity - tolerant of the commercially important tilapia species, but grows relatively well in salinities up to 15 ppt. It has been reported that Nile tilapia can reproduce at salinities up to 10 - 15 ppt, but the reproductive performance is better at salinities below 5 ppt.

A Global Problem: A Fish-Born Solution

Increasing protein production from aquatic resources has a high potential of closing the gap between animal protein demand and supply. The high potential of fish production in this regard emerges from the fact that they occupy niches that are not commonly utilized for producing other conventional items for human consumption. Another important advantage is that many fish species are low on the food chain, and, hence, may provide a two-sided solution that helps closing the nutritional gap while contributing to solving problems of bioaccumulation of wastes in natural water. Natural fisheries are now utilized at a maximum. Very little room is available for increasing fish landing. Aquaculture is the most viable option to continue attempts to fill the global nutritional gap and to alleviate the enormous pressure on natural fisheries that are currently exploited at a maximum with very little room for increase in fish landings. While contributing to food security and ridding the environment of biological waste, fish become less expensive to grow than most other sources of animal protein. Because their feed conversion ratio is generally lower than large animals and poultry, their cost of production may be brought further down and they may exert less pressure on available food resources. Aquaculture may exert additional demand on water resources that are increasingly becoming at high demand. The Nile is the major source of fresh water in Egypt. Egypt's share of the Nile water is currently exploited completely. Any attempts to increase agricultural or aquacultural production will mean the need for additional water resources. With better management and additional research, solutions to this problem can be provided through increasing water productivity. Possible means for this include different forms and levels of integration as well as maximization of harvest per unit of both land area and water volume.

TILAPIA, AN ECONOMICALLY AND BIOLOGICALLY, FEASIBLE INVESTMENT

Nile tilapia is a widely accepted fish in many markets around the world. Its suitability to culture in fresh water ponds, cages, and raceways make it a species of choice for fish farmers as well. Selection for improved body weight has proven success in many terrestrial and aquatic species including Nile tilapia. The GIFT strain of Nile tilapia was developed in the Philippines by The WorldFish Center through its project on the genetic improvement of farmed tilapia.

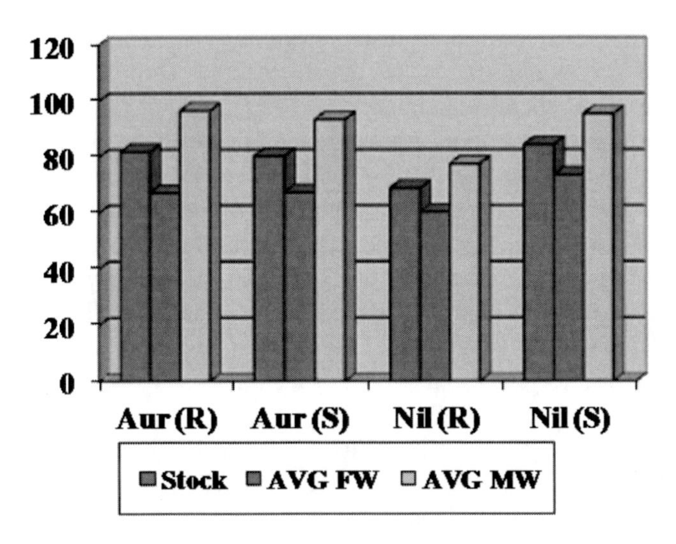

Figure 2. Improvement after two generations of mass selection in Nile tilapia, *O. niloticus*, at private commercial tilapia farm, Egypt.

The recent selected generations of the GIFT tilapia outperform the parental populations by more than 80%, indicating a high potential for achieving similar levels of improvement in Africa. Improved Nile tilapia is now being used in aquaculture in many Asian countries. These efforts may have their most significant impact in Egypt where fish production is relatively high and the contribution of aquaculture to fish production and, hence, to the national economy is immense. Selective breeding work carried out by the author of this chapter in a private commercial tilapia farm in Egypt resulted in an improvement of 23% in tilapia body weight relative to the control after two generations of mass selection (Figure 2).

GENETIC IMPROVEMENT … CAN THE POTENTIAL BE MAXIMIZED?

Genetic gain from selection programs can be maximized if the procedure could be continued throughout the year. In Egypt, the seasonal nature of the climate limits the effective months for working in the field environment to the 6 months of May through October. Working with a species like *O. niloticus* that matures at 6 months of age, such climatic conditions would limit the number of selection cycles to 1 cycle/year, a limitation that inflects its negative effect on the gain accomplished from the selection program within a given period of time. The aim of the selection program is the maximization of productivity from the ongoing family/within family selection program through ability to produce families and test them under intensive culture conditions and to continue the selection and evaluation procedures at times when outdoor conditions are not favorable. This can be achieved through conducting the selection work indoors where temperature and other environmental variables can be monitored and controlled. As the improved fish are tested under commercial aquaculture conditions, it is expected that the cumulative improvement through doubling the number of generations produced within a specified period in addition to the high correlation between fish performance in the selection environment and the testing environment will outweigh the improvement achieved through selection in the testing environment directly. At the same time, an improved stock that is highly fit for intensive culture will be available for commercial aquaculture in Egypt which is steadily moving toward intensification. Moreover, since the environmental conditions in the system will be controllable, the system can be used to mimic the conditions in a given African country and evaluate/develop useful strains for use in that particular country. The selection program includes a set of experiments that are designed to produce the maximum possible improvement under experimental conditions, to measure correlation between performance in the selection environment and performance under other environmental conditions, to measure the response to selection for increased body weight, and to monitor some of the correlated responses that are important either from an aquacultural or from an economic point of view.

Selective Breeding

Selection is a breeding technique in which superior individuals in a population are kept and allowed to mate to produce offspring that is superior to the original population. It becomes feasible when a significant portion of the phenotypic variance is due to the additive component.

This is a strategy to utilize the "additive genetic variability" in a population by choosing the "best" individuals as breeders. Selective breeding can be used to improve several traits simultaneously, and improves the traits by moving the average performance of individuals in the breeding population.

Although there are many technologies available for genetic enhancement of fish, only selective breeding can assure a long-term genetic enhancement of fish (Figure 3). Therefore, other technologies should not be looked upon as alternatives, but rather as additional technologies to selective breeding. Biotechnology (DNA-markers, gene-mapping etc.) can be used to improve breeding programs.

The ratio between these two quantities is called heritability, h^2. The higher the heritability of a trait the easier and faster it is to improve that trait by selection. Heritability values less than 0.2 are considered low, those between 0.2 and 0.4 are moderate, and above 0.4 are considered high. Traits with low heritability values are hard to improve by selection. The most realistic way to measure heritability is the realized heritability which is expressed as the response to selection divided by the selection differential. Heritability values for growth rate are moderate to high in channel catfish and several salmonid species. Hence, selection programs were successful in these species. All attempts to improve growth rate in *Oreochromis niloticus* through selection failed because of problems with low heritability. Tracing the populations that were used in these experiments showed that they were small ones or they ran through a bottleneck at some point in their history. This should result in deleterious inbreeding effects. When the heritability is low, certain more efficient selection techniques can be used. Family selection, e.g is about 4.7 times more efficient than individual selection. I currently carry out attempts to improve growth rate in Nile tilapia through both mass selection and family selection using a broad genetic background in the base population. Another program for QTL mapping and marker assisted selection is underway in collaboration with different research institutes.

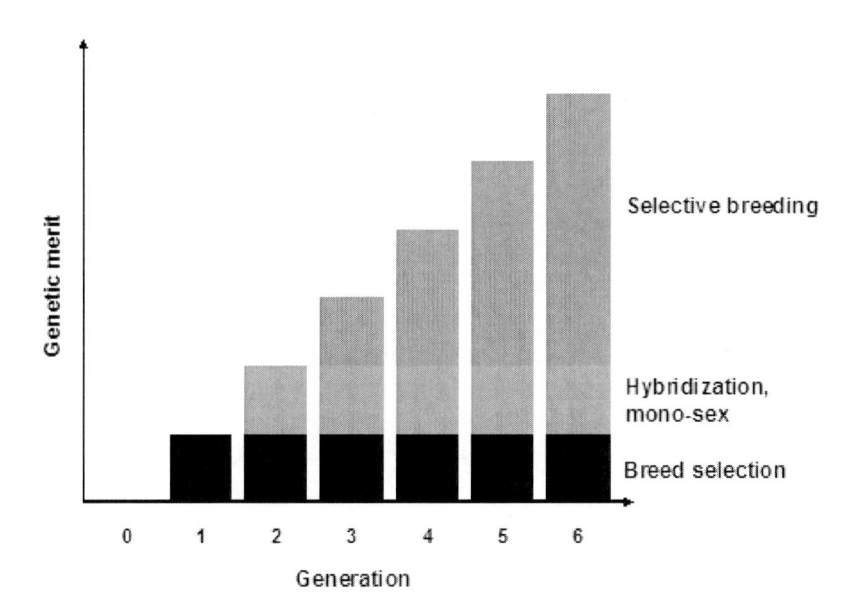

Figure 3. Selective breeding is the only technology that can provide a continuous improvement of fish.

Overview of Tilapia Genetic Improvement Programs

The production of Nile tilapia is expanding fast in many countries around the world, the production has increased from 127,000 t in 1988 to 2.3 million tons in 2008, making tilapia the most important aquaculture species in the world [6; 7]. Tilapia farming has received an increasing attention around the world during the past three decades [8; 9; 10; 11; 12; 13; 14; 15; 16; 17; 7; 18] and the most widely farmed fish [19], with farming in more than 100 countries in 2002 [15] and has led to the adoption of several breeding programs for improving its productivity [1; 2; 34; 5].

Increasing the productivity from aquaculture of African fish, especially tilapia, may help in encouraging the adoption of aquaculture as a means of increasing animal protein production. Tilapia culture can contribute significantly to solving the problem of animal protein deficiency, especially in the third world.

Fish have fewer resources to compete for with other conventional agriculture and livestock products [20; 21; 22; 23; 24; 25]. Nile tilapia is an excellent fish for aquaculture that has a proven potential for improved production through selective breeding both in the Philippines, the GIFT strain, and in Egypt, the Abbassa strain [26]. The Gift tilapia was produced from a composite population that originated mainly from two Egyptian and one Kenyan strains of Nile tilapia. The GIFT project has had a world-wide impact on aquaculture.

In 2005, Nile tilapia represented 84% of total tilapia production compared with 42% in 1988 [27]. The remarkable success of the elaborate genetic selection technique that was utilized in developing this high performing strain for commercial aquaculture in the Philippines and Southeast Asia should encourage attempts to develop improved African strains for aquaculture from locally available wild stocks [26; 28; 21; 29; 27]. The Abbassa strain of genetically improved Nile tilapia was developed along the same lines while applying similar technologies to native Egyptian Nile tilapia stocks. Although there are many technologies available for genetic enhancement of fish, only selective breeding can assure a long-term genetic enhancement of fish.

Applied Selective Breeding Programs of Tilapias in Egypt

Mass Selection

Mass selection is utilized for its simplicity and applicability with limited requirement for expertise and/or sophisticated analysis. The largest 10% females and males are selected and used as brood fish for the next generation (Figure 4).

A random population is also maintained for comparative evaluation. Growth of experimental fish is monitored through monthly samples during warm months. An improvement of 14% of the selected population over the random one of *O. aureus* has also appeared in the third generation. Brood fish from both species were selected and reproduced for production of the 4th generation.

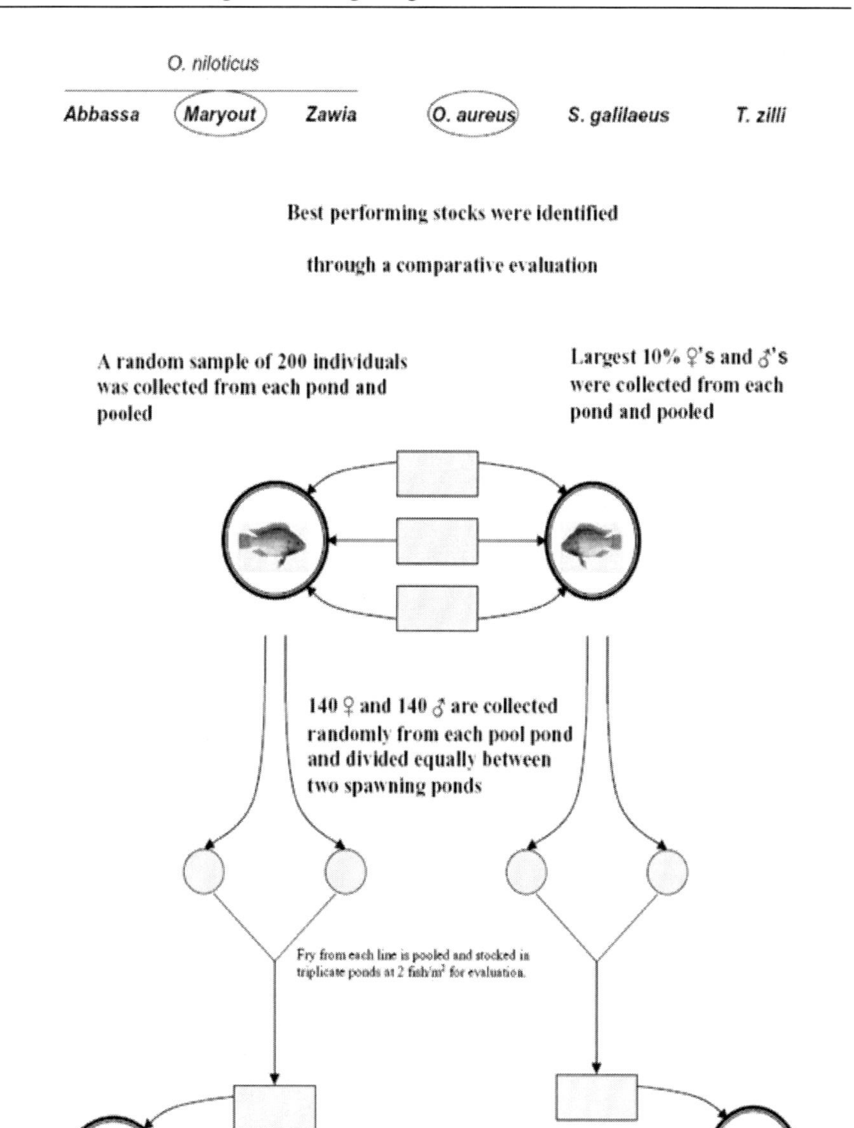

Figure 4. Diagrammatic representation of the procedure involved in mass selection of both *O. niloticus* and *O. aureus.*

Mass Selection for Increased Body Weight

Three populations of Nile tilapia, *Oreochromis niloticus*, and one population of each blue tilapia, *O. aureus*, white tilapia, *Sarotherodon galilaeus*, and green tilapia, *Tilapia zilli*, populations have been used to initiate this project. These populations are: -

- *Oreochromis niloticus*, Abbassa population: Collected in spring of 1999 from the Ismailia Canal East of the Nile Delta near Abbassa, Egypt.
- *O. niloticus,* Maryout population: Obtained in summer of 1999 from a hatchery near Lake Maryout, a brackish water lake in the northwestern part of Egypt south of Alexandria.
- *O. niloticus*, Zawia population: Collected from El-Zawia region north of the Nile Delta near Lake Burullus in summer of 1995. They represented wild fish that accidentally entered the ponds of a governmental fish farm while filling the ponds with water or while stocking the ponds with mugilid and cyprinid fry.
- *Oreochromis aureus*, *Sarotherodon galilaeus*, and *Tilapia zillii* were collected from Ismailia Canal in spring of 1998. These were wild fish that originally entered from, and then through the internal supply canals.

Maryout and Zawia populations of *O. niloticus* were derived from about 300 adults each. Fish were reproduced using equal numbers of females and males. A random sample of their progeny was used as brood fish. All other populations were derived from 400-500 adults each. Brood fish from each population were reproduced in two 100-m^2 circular spawning ponds at the the farm facilities during the last week of June 1998. Stocking density was one individual/m^2 at a sex ratio of 1:1 to maximize the effective population size. Ponds were drained 35 days after stocking and fry were collected, counted, and held in 2 m^3 hapas to recover from handling stress. Fry from each group were counted, weighed and stocked into 3 randomly assigned 1000 m^2 earthen ponds at a density of 20,000/ha. Because of low fry production *by S. galilaeus*, three 100-m2 ponds were used to maintain the same stocking density as in other groups. Fish were fed daily with 25% protein sinking fish feed manufactured by the Alexandria Oil and Soap Company, Kafr El-Sheikh, Egypt. Feed was delivered about midday on 70 X 70-cm wooden feeding trays that had a 20 cm rim attached to the tray at a slope of ½. Trays were examined

1 hr later and feeding rates were adjusted according to consumption. Water depth in the ponds was about 1 m. Dissolved oxygen, pH, and Secchi disk visibility were measured daily, but remained at acceptable levels. Temperature was recorded hourly at 5, 35, and 70 cm below water surface using a Campbell Scientific CR10 Measurement and Control Module. Daily maximum and minimum temperatures were used to calculate an average temperature for the periods that preceded each sample. Fish were harvested starting December 20, 1998 and sorted by sex. Fish were counted and weighed at harvest. Their performance was evaluated concerning several important traits including growth performance, sexual dimorphism, survival, feed conversion ratio, and seinability. The Maryout population of Nile tilapia and the blue tilapia population were identified as the best performers in most of the traits that were studied, including individual growth. These two populations were therefore used as base populations for mass selection programs. Random sample of 200 individuals were collected from each of the 3 replicate ponds for each species and weighed. Samples collected from all three ponds were mixed in a common pond. Female and male proportions in the samples were adjusted to their proportions in the pond they were collected from. The heaviest 10% females and males were selected from each experimental pond, weighed, and pooled in another pond. Starting 1999 until present, the two lines, random and selected, of each species have been maintained so that a random sample is collected from the random line and the heaviest 10% males and females are selected from the select line in each generation as described below (Figure 4) and harvesting has been shifted to beginning of April of each year to extend the winter testing period and avoid possible handling mortality during winter harvesting.

A Preliminary Evaluation of Cold Tolerance in Mass Selected and Unselected Nile Tilapia, *O. Niloticus*, Fry

A sample of selected and control Nile tilapia, *O. niloticus*, fry produced in July 2003 was collected and counted. Five hundred individuals from each line were transferred to a 6 m^2 hapa. Both hapas were installed in one earthen pond and fed with isocaloric 40% protein diet. Fry were reared in the hapas to an average weight of 2 g. They were then passed through a grader to ensure size similarity between the two lines. Fry from each line was then transferred to a cold room served with a thermostatically controlled chiller where they were randomly assigned to one of two glass aquaria, 450 L each, at a rate of 100 fry/

aquarium. The water in the aquaria was adjusted to equal temperature in the pond, 25°C. No feeding was provided in the aquaria. Continuous aeration was provided through air stones connected to blowers. Faeces were siphoned out of the aquaria daily and water replaced with clean water at the same temperature. Fry were allowed 24 hours to settle in the aquaria. Temperature dropped to 21°C. The chiller was then used to gradually reduce the temperature to 16°C then to 13°C over periods of 48 hours each. The cold challenge was then started by reducing the temperature one degree every 24 hours. Temperature and fish mortality were monitored on hourly basis and recorded.

Reproductive Behavior and Inbreeding under Mass Spawning Conditions in Nile Tilapia, *O. Niloticus*

Nile tilapia brood fish were individually tagged and stocked in two hapas 16 m^2 each. 24 males and 50 females were equally divided between two hapas. The hapas were checked twice a week for spawning. Eggs were removed from the mouth of incubating females, transferred to the laboratory for hatching, and the resulting fry were reared in aquaria. DNA was collected from fry and parents for microsatellite analysis. Data will be used to reconstruct the pedigree of the fry and determine participation in the mating by each fish and the potential for inbreeding.

Progress to Date

Mass Selection for Increased Body Weight

Four generations of selection in *O. niloticus* and *O. aureus* have resulted in 35% and 14% improvement, respectively. Mean individual weight at harvest (s.e.) in grams for random and selected O. niloticus was 78.2 (2.4) and 105.3 (17.5), respectively. These weights in *O. aureus* were 89.1 (13.6) and 101.4 (6.8), respectively. Genetic improvement in *O. niloticus* proceeded faster than that in *O. aureus* (Figures 6 and 6). Each stock has been evaluated in separate ponds. Pond differences may have contributed to the different magnitudes of standard errors. Tilapia is a warm water fish that can be grown only during the warmer summer months in Egypt. Since the improvement achieved in this project is obtained under low temperature conditions (in winter), it is expected that the farmers may either be able to extend the growing season and make use of the colder months of the year to produce larger fish or split the extended growing period making use of the accelerated growth to produce two crops per year.

A Preliminary Evaluation of Cold Tolerance in Mass Selected and Unselected Nile Tilapia, *O. Niloticus*, Fry

Initial experiments conducted on cold tolerance in selected and control tilapia fry showed no difference between performance of the random and the selected lines. Mortality started at 11°C in both lines. Complete mortality was reached in the selected and random lines at 9°C and 8°C, respectively.

DELIVERY OF RESULTS (IMPACT PATHWAY), ANTICIPATED BENEFITS TO FARMERS

The experiments we are conducting to better understand reproductive behavior under mass spawning conditions are expected to help develop a clearer picture on proportion of males and females participating in mating, the frequency of repeat spawning, and the deviation from random mating. Field studies to evaluate the impact of using improved seed on farm productivity and economics are currently underway in Egypt. A field trial is currently underway in a commercial farm in which two ponds, 0.4 ha each are stocked with progeny of the 4th generation of selection and are compared to two ponds stocked with the commercial fry used by the farmer.

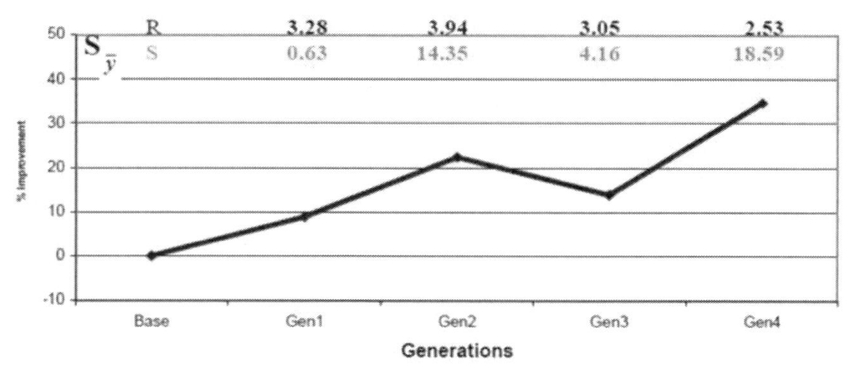

Figure 5. Improvement in mass-selected Nile tilapia, *O. niloticus*, over the unselected Control obtained through four generations of selection.

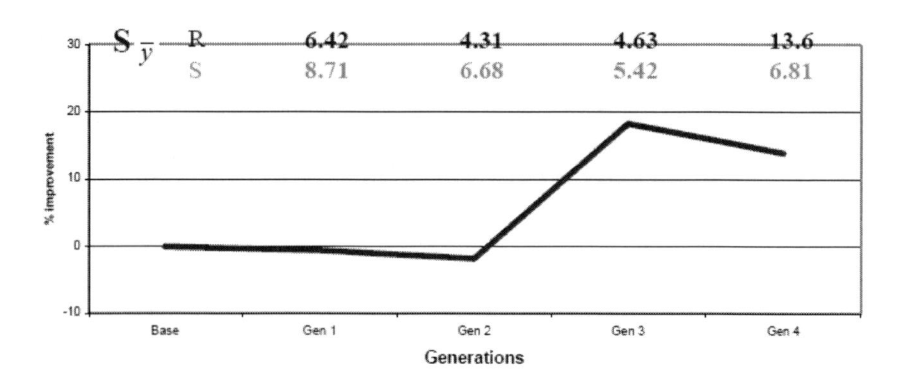

Figure 6. Improvement in mass-selected Nile tilapia, *O. aureus*, over the unselected Control obtained through four generations of selection.

Pedigree Selection

Populations Sampled and Schedule of Matings

Selection

Four stocks of Nile tilapia, *Oreochromis niloticus*, were collected from different geographical locations within Egypt between 1995 and 1999 (Figure 7). 0.5 mm mesh plastic was used to construct 120 hapas 1.5X1.0X1.0 m that were installed in a single 4000 m^2 earthen pond and used for reproduction and fry maintenance. Brood fish from the four stocks were chosen based on the absence of apparent signs of diseases and/or deformities and separated by sex for three weeks. They were then paired in summer 2000 for reproduction in a complete 4 X 4 diallele crossing. A total of 72 pairs were formed with equal representation in all possible mating combinations. Checking for spawning was conducted weekly and occurrences of spawning recorded. If a pair did not reproduce, the female was replaced after two weeks. If the lack of reproduction continued, the male was replaced after another two weeks. According to initial calculations, the plan was to terminate this phase of spawning when 70% of the pairs spawn (about 50). Spawned males were mated again with new females. The two females that were mated with a given males were always from different stock. The 2nd phase was to be terminated as 70% of the pairs reproduce (35 pairs). The above process led to the production of 37 half-sibs (HS) groups (males that reproduced with two females each) while 25 males reproduced with one female only. The rest did not spawn at all.

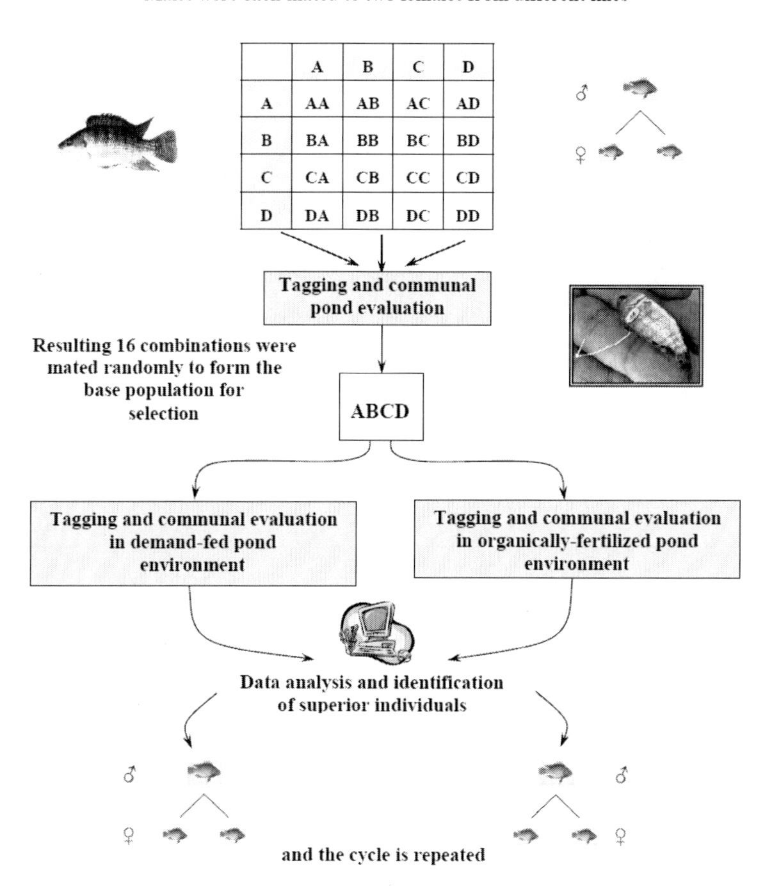

Figure 7. Diagrammatic representation of the procedure involved in pedigree selection of both *O. niloticus*.

Following reproduction, each group of full sibs was reared in its respective hapa and fed with powdered formulated feed containing 40% protein. Fish were reared to an average weight of 0.5 to 1.0 g. They were then thinned to 50/hapa and grown again to an average weight of 2.0 g before tagging. Fish were then tagged using fingerling Floy tags, Floy Tag Company, USA, and returned to their respective hapas for two days to recover from tagging stress. Because the reproduction process took 12 weeks, progeny of males that failed to reproduce a second time were excluded from the experiment in hope to enhance reproductive activity in future generations of the selection program (Table 1). A total of thirty fish were selected at random from each full sib group, their weight and standard length were recorded and all groups were stocked in a 1000 m^2 earthen pond on 30 August for communal evaluation. Fish werefed with formulated 25% protein feed until they were harvested on 1 April 2001, individually weighed and their standard length and body depth were measured. A total of 1286 individuals were recovered from a total of 2100 that were initially stocked.

Table 1. Summary of calculations for the different lines in the pedigree selection

Spawning season	Line	Sires	Dams	Progeny
2000	Diallele	37	74	2100
2002	Base (high input)	84	112	2130
	Base (low input)	80*	105*	2120
2003	High input	47	86	3450
	Low input	50	86	2087
2004	High input	58	98	?
	Control	22	22	?
Total		276*	456*	11887

* Sires and dams appearing in the Base (low input) category in 2002 were a subset of the ones appearing in the Base (high input) and, hence, were not included in the total.

Data was analyzed using the ASREML software. Heritability estimates and the associated standard errors for these traits were obtained.

After harvesting, the resulting 16 crosses were used in a random mating in spring 2002 with the only restriction that full- and half-sibs were not allowed to mate with each other. This led to the production of a mixture of pure bred, two-, three-, and four-way combinations in the progeny. The resulting synthetic breed was used as a base population for selection. From this generation on, each male was placed in a hapa with two females. Spawning was checked twice a week. Whenever a female reproduced, she was left in the hapa with her progeny. The male and the other female were transferred to a new hapa to reproduce. Progeny of males that reproduced only once were used in the communal evaluation. Sixty individuals from each Full-sib group were tagged with fingerling Floy tags. Tagged fish were divided into two groups that were all stocked in earthen ponds. The first group was stocked at 2 fish/m^2 and fed ad libitum with 25% protein sinking feed while the other was stocked at 1 fish/m2 in ponds fertilized with chicken manure. Fish feeding and manuring treatments were isonitrogenous. To ensure equal numbers of fish being used in both lines of selection, equal selection intensity and similar rates of inbreeding, two fertilized ponds and one fed pond were used in the study. Final weights of the fish were recorded and analyzed using pedigree information. Based on the estimated breeding values, best performing males and females were selected in 2003 and used as brood fish for the second generation (Table 2).

Table 2. The annual schedule of events

Activities	Spawning season
Mating	End of May-June
Fry nursing/rearing	June-July
Tagging	July
Grow-out	August-April
Harvest	1st week of April
Data entry/analysis	April

Progress to Date (e.g., No. Families Generated, Genetic Evaluation, Selection Response, Achievements and Constraints, Problems if any and Proposed Solutions)

Heritability estimates (s.e.) for body weight, standard length, and body width at harvest in April 2001 obtained in the diallele cross were 0.32 (0.16), 0.30 (0.15), and 0.16 (0.12), respectively. Based on these results, it was determined that direct measurement of body weight is the easiest and most practical way of measuring and selecting for growth. Heritability estimates for body weight in the base population (2003) under high and low input conditions were 0.26 and 0.27, respectively. These estimates indicate the potential for improving body weight through selection under both low input and high input conditions since one fourth to one third of the variability in body weight oh the fish is due to additive genetic effects. Selection response per generation will be measured in the next generation (2004) using a comparative performance evaluation of the selected individuals with the control population (Table 3).

Table 3. Descriptive statistics in the selection of high and low input lines

Variable	Sex	N	Mean	Minimum	Maximum	Standard Deviation	Coefficient Variation (%)
High input line 2003	Female	544	48.41	13.2	158.2	25.9	54
	Male	442	77.01	14.0	229.0	45.6	59
Low input line 2003							
High input line 2004	Female	915	40.9	8.0	153.6	18.6	45
	Male	939	65.8	8.8	175.7	29.4	44.6
Low input line 2004	Female	599	79.6	18	169.6	25.5	32
	Male	684	101.6	23.8	214.8	34.1	34

Delivery of Results (Impact Pathway), Anticipated Benefits to Farmers

The high input line being used in a field trial. Results of this study along with the results of the GXE experiment where performance of the high and the low input lines are evaluated in a range of environments will help in determining the future of both lines and their potential benefits to farmers. Selection in pedigreed lines allows for control of the mating structure and, hence, inbreeding. This usually results in higher efficiency and longevity of the breeding program. Depending upon the results, the program is planned to continue for 10 years.

Combined Family Selection

All possible two-way-cross combinations between 4 populations of Nile tilapia were randomly paired in 2006 to produce the base population for the combined family/within family selection program. To be able to start the breeding season early, breeding was conducted in heated and covered concrete tanks. Full- and Half-sib families have been produced and fish were individually tagged.

Selection under Commercial Aquaculture Conditions

In this selection program, communal evaluation of full-and half-sib families is conducted under commercial farming conditions in Egypt. Based on data analysis, males and females with the highest breeding values will be selected and used for producing the next generation.

Selection under Low Input Conditions

Individuals from the same base population are communally evaluated in low input, i.e., fertilized-pond, conditions. Comparative performance of both selected lines will be studied in reciprocal environments after two generations of selection. An experiment on fry performance in low input systems was conducted. An attempt was made toward synchronization of spawning in Nile tilapia.

Marker-Assisted Selection in *O. Niloticus*

Performance of the 4-way crosses of Nile tilapia has been evaluated in ponds. Sample collection was more elaborate than originally planned. Sexual maturity, tolerance to low dissolved oxygen and high ammonia were examined, Blood samples were collected from individuals with extreme breeding values for growth in addition to all above tests.

CONSTRAINTS AND DIFFICULTIES ENCOUNTERED

Mass Selection

Unexpected severe cold winter temperatures caused a differential mortality in the two populations of *O. niloticus*, 34% in the random line vs. 16% in the selected one. The difference in fish density in the culture ponds may have affected growth and masked the expected response to selection. *O. aureus* is more resistant to cold than *O. niloticus* and, hence, threshold temperatures for Nile tilapia did not affect their survival.

Family/within Family Selection

Loss within brood fish prior to the breeding season and the unexpected increase in aggressive behavior during the breeding season was the two major constraints. These were resolved with rapid conditioning and utilization of backup stocks with the same genetic backgrounds. Delay in transferring experimental fish for the technology transfer project affected availability of facilities for indoor spawning of brood fish. The above difficulties equally impacted the family/within family selection program in low input farming systems and, to a lesser extent, the marker assisted selection work. The lack of a reliable recirculating system affected the spawning synchronization experiment.

Delivery of Results (Impact Pathway), Anticipated Benefits to Farmers

The high input line being used in a field. Results of this study along with the results of the GXE experiment expected in 2004 where performance of the high and the low input lines are evaluated in a range of environments will help in determining the future of both lines and their potential benefits to farmers. Selection in pedigreed lines allows for control of the mating structure and, hence, inbreeding.

This usually results in higher efficiency and longevity of the breeding program. Depending upon the results, the program is planned to continue for 10 years. Tables (4, 5, 6, 7, 8, 9, 10 and 11) represent the presentation of the data collected from the pedigree selection program.

Table 4. Number of sires, dams and progeny, by spawning season and line

Spawning Season	Line	Sires	Dams	Progeny
2002	Base Population	86	115	986
2003	Selection	48	86	1847
2004	Selection	59	101	1088
	Control	21	21	250
Total		214	323	4171

Table 5. Schedule of reproduction and management

Activities	Spawning Season		
	2002	2003	2004
Mating	Jun - Jul 2002	Jun - Jul 2002	Jun - Jul 2004
Nursing Hapas	Jun - Jul 2002	Jun - Jul 2002	Jun - Jul 2004
Rearing Hapas	Jul - Aug 2002	Jul - Sep 2003	Jul - Aug 2004
Tagging	Aug 2002	Sep 2003	Aug 2004
Grow-out	Sep 2002 – Apr 2003	Sep 2003 – Apr 2004	Sep 2004 – Apr 2005
Harvest	7 Apr 2003	5 - 7 Apr 2004	16 Apr 2005

Table 6. Number of observations (N), simple mean, minimum and maximum, standard deviation and coefficient variation and standard deviation of initial weight (g), harvest weight (g), age (days) at harvesting and survival

Variable	N	Mean	Minimum	Maximum	Standard Deviation	Coefficient Variation (%)
Initial Weight	7482	2.9	0.3	66.0	2.5	86.6
Harvest Weight	4170	73.6	8.8	272.9	46.6	63.3
Age at Harvesting	4170	292.9	259.0	356.0	18.2	6.2
Survival	9267	0.45	0.0	1.0	0.50	111

Note: The descriptive statistics in Table 6 were obtained from SAS by using Proc Means. The coefficient of variation was very high for initial weight and harvest weight. Initial and harvest weight were transformed to logarithm (base 10) for all the later analysis.

Table 7. The sex ratio in different spawning seasons

Spawning Season	N		Male : Female
	Male	Female	
2002	440	546	1 : 1.24
2003	938	909	1 : 0.97
2004	655	682	1 : 1.04

Table 8. Analysis of variance of initial and harvest weight and survival: Tests of fixed effects using PROC MIXED

	Initial Weight		Harvest Weight		Survival	
Effect	F Value	Prob. > F	F Value	Prob. > F	F Value	Prob. > F
SS x E x L x S*			0.63	0.8342		
SS x E x L*	23.17	< 0.0001			48.59	< 0.0001
Age at harvest (SS, E, L, S)			3.51	< 0.0001		
Residual Variance	0.0256		0.0243		0.2183	

* SS = spawning season; E = environment; L = line; S = sex.

Note: Table 8 shows the level of significance for the corresponding subclasses fitted as fixed effects and for the linear covariate that we fitted in the analysis using SAS. Initially the fixed effects were fitted separately, as well as two way interactions. Two way interactions were mainly non-significant, and hence the subclasses described in Table 9 were fitted for the different traits. For initial weight and survival, sex was not fitted in the model. This is because the sexes were not fully available for initial weight (only those survived after harvest will have the sexes). For survival is the same as the initial weight, only those are survived will have the sex identified.

Table 9. Least squares means for the different levels of the fixed effects treatment and sex

Variable	Effect		Least Squares Means	Standard Error
Initial Weight	Spawning Season	2002	0.2123[a]	0.0352
		2003	0.5203[bc]	0.0451
		2004	0.2195[ac]	0.0273
	Line	C	0.2388[a]	0.0532
		S	0.3960[b]	0.163
Harvest Weight	Spawning Season	2002	1.6962[a]	0.0226
		2003	1.6409[b]	0.0237
		2004	1.9833[c]	0.0175
	Line	C	1.7468	0.0339
		S	1.8001	0.0088
	Sex	F	1.6825[a]	0.0176
		M	1.8645[b]	0.0177
Survival	Spawning Season	2002	0.4891[a]	0.0261
		2003	0.5496[b]	0.0268
		2004	0.3784[c]	0.0197
	Line	C	0.4942	0.0381
		S	0.4506	0.0101

Means without a common superscript are significantly different ($p<0.05$).

Note: The adjusted means were obtained by using Lsmeans (least square means) option in Proc Mixed is shown in Table 9 (log10 transformed). From the results, we can see that the control and selected line were not statistically significant, except initial weight. There were significant differences among spawning seasons and between sexes.

Table 10. Variance components, heritability and maternal common environment effect for initial and harvest weight, and survival

Parameter	REML Estimate		
	Initial Weight	Harvest Weight	Survival
Additive genetic Variance (σ^2_A)	0.010	0.009	0.029
Maternal & Common Environment Variance ($\sigma^2_D = \sigma_{M\_Ec}{}^2$)	0.062	0.024	0.004
Phenotypic Variance (σ^2_P)	0.095	0.064	0.244
Heritability (standard error) [h^2(s.e.)]	0.106 (0.0576)	0.144 (0.0598)	0.120 (0.0346)
Maternal Common Environment (standard error) [C^2(s.e.)]	0.652 (0.0475)	0.384 (0.0496)	0.015 (0.0214)
Initial Weight		0.638 (0.0210)	0.083 (0.0196)
Harvest Weight	0.233 (0.3148)		0.420 (0.0177)
Survival	0.008 (0.2558)	0.458 (0.2035)	
Initial Weight			
Harvest Weight	0.843 (0.0408)		
Survival	0.378 (0.3710)	0.691 (0.4581)	

Above the diagonal - Phenotypic correlation (standard error).

Below the diagonal - Genetic correlations (standard error).

Below the 2nd diagonal - Common environment correlations (standard error).

Note: The phenotypic and genetic parameters were estimated by using ASReml. I am using a trivariate analysis (initial weight, harvest weight and survival; both live weight were logarithm (base10) transformed) to better estimate the above parameters. In this analysis, different traits have different fixed effects, i.e., for initial weight and survival, the fixed effect is the interaction between generation, environment and line and subclasses only, and the fixed effects for harvest weight are interaction between generation, environment, line and sex and subclasses, with age at harvest as covariate within generation, environment, line and sex sub-class. In this trivariate analysis, spline was only fitted to age at harvest for harvest weight (only this trait has age at harvest) as one of the random effects. Animal and dam (genetic effect) were fitted as random effects too.

Table 11. Response to selection estimated by different methods and expressed in different ways

Method		Model (effects)	Trait	Selection Response[A]		
				Actual Units	Percentage	Genetic Standard Deviation Units (Actual/σ_A)
i.	Comparing the estimated breeding values between the progeny of the 2002 spawning season and those of the Selection line in the 2003 spawning season.	Fixed: SS x E x L x S[*] Covariate: Age at harvest (SS, E, L, S) Random: Spline(age at harvest), animal, dam	Initial Weight	0.007	1.67	0.070
			Harvest Weight	0.052	2.88	0.547
			Survival	0.040	8.80	0.235
	Comparing the estimated breeding values between the Selection line in the progeny of 2003 spawning season and those of the Selection line in the 2004 spawning season.		Initial Weight	0.013	3.18	0.130
			Harvest Weight	0.054	3.02	0.568
			Survival	0.039	8.63	0.229
ii.	Comparing the estimated breeding values of the Selection and Control lines in progeny of the 2004 spawning season.	Fixed: SS x E x L x S Covariate: Age at harvest (SS, E, L, S) Random: Spline (age at harvest), animal, dam	Initial Weight	0.010	2.65	0.100
			Harvest Weight	0.030	1.66	0.316
			Survival	-0.007	-1.55	-0.041

[A] Actual units are LW or survival difference in mean breeding values for methods (i) and (ii); Percentage is the least squares means relative to mean of LW and survival for the selection population (for initial weight = 0.396, for harvest weight = 1.800, for survival = 0.451); genetic standard deviation equals the square root of the additive genetic variance in Table 5 ($\sigma_{A(initial)} = 0.100$, $\sigma_{A(harvest)} = 0.095$, $\sigma_{A(survival)} = 0.170$).

[*] SS = spawning season; E = environment; L = line; S = sex.

Note: The results in Table 11 were also estimated by using the same model as mentioned in above section (Table 10). Table 16 shows the selection response between generations (method i) and lines (method ii) by comparing the estimate breeding value.

Relative Efficiency of Selection Based on Estimated Breeding Values Using Information from One or Two Generations

For generation 2 I calculated EBV base on the individual's information, and that of its full and half sibs within a generation. We also estimated the EBV for the same individuals using pedigree and performance information from that and from the previous generation. The simple correlation between BVs calculated in the two different ways was 0.76. This indicated that I lost 24% of efficiency in selection if I accept that the estimates using both generations' information and the pedigree are better.

Relative Efficiency of Selection Based on Estimated Breeding Values Using Information from One or Three Generations

Similarly, for generation 3 I calculated EBV based on the individual's information, and that of its full and half sibs within a generation. I also estimated the EBV for the same individuals using pedigree and performance information from the previous two generations as well as that from the current generation. The simple correlation between EBVs was 0.65; this indicated that we lost 35% of efficiency in selection by not using the pedigree and performance information. Compared to the previous analysis with only two generations, the lost of efficiency is now greater. This is because with three generations together, I have more to gain and more information for estimation using all the pedigree, so when I ignore the pedigree information the loss of efficiency becomes greater.

Field Trials

The purpose of this study was to obtain estimates of some quantitative genetic parameters aiming at the estimation of breeding values of individuals in the population based on available pedigree information in each generation leading to accurate identification of selection candidates and maximizing the genetic gain.

Preliminary Report About Results of the Field Trial on the Mass Selected Nile Tilapia Conducted at Commercial Farm

The farm was inspected and four empty ponds, two acres each, were investigated and assigned for the experiment in agreement with the farmer. Brood fish were transferred to the hatchery 2-June-2004 for simultaneous production of fry from both the selected and the commercial lines. Empty ponds were investigated before stocking and seepage water was pumped out Weeds were cut and removed. Pond bottoms were disinfected by spraying chlorine to eradicate unwanted fish. Fry transfer to the farm ponds took place on 26-June-2004. Two ponds were assigned for each line at random and stocked with fry at 2 fish/m^2. Ponds were fed at 3% of their body weight with 25%protein feed. Periodical sampling (Table 12) was used as an indication of growth rate and as a basis for calculating feed quantities for each pond.

Table 12. Sampling dates and average weights of fish in commercial and selected fish ponds at a commercial tilapia farm

Date	Average weight (g)		Growth difference (g)
	Commercial Line	Select Line	
26 June (at stocking)	0.17	0.16	-0.01
12 Aug	30.0	29.0	-1.0
01 Sep	43.0	48.5	5.5
16 Sep	50.0	77.0	27
30 Sep	68.5	89.0	20.5
15 Oct	83.0	112.5	29.5

Sampling was then stopped as water temperature got lower and fish seining became more difficult. Based on the mean weights in the last sample, it appears that there was an improvement of 36% in body weight of the selected fish over the commercial line.

Average daily temperature in January and February 2005 was colder than usual. Minimum air temperature recorded at Nile Delta was as low as 3° C. This was reflected in pond water temperature as well. Water temperature dropped to a low of 10° C and remained at lows that were critical for tilapia survival (below 12° C over two periods. One was between 17 - 28 Jan and the other was between 6 - 17 Feb. This led to unexpected fish mortality in the experimental ponds during February. Several visits to the farm were arranged to monitor and assess the situation. It was decided that treating the fish would be very difficult. Because the cold weather continued and mortality did not stop, it was decided later on to terminate the experiment and salvage the situation. It was not possible to get an accurate tally on the mortality because the dead fish from different ponds were occasionally mixed together during pick up from the ponds and because the bird and dog intervention in collection of dead fish were difficult to control in a large farm. However, we were able to get good estimates based on the stocking and harvesting data.

Breeding Program for Fillet Trait in Nile Tilapia

Aquaculture become the promising source for increasing protein production from aquatic resources and has a high potential of closing the gap between animal protein demand and supply. In the market, tilapias are sold as either whole fish or as fresh or frozen fillets. The US and European markets prefer fillet [2; 3; 29; 7]. The type of product determine the fish size to be produced; for whole fish preferable 400–500 g, which can be attained at the age of 8 to 10 months, while for fillet 700 to 1000 g is preferred (11 to 14 months) [19]. Fillet yield vary depends on different fish species, for Nile tilapia (*Oreochromis niloticus*, 34–37%) [29; 2] for rainbow trout (*Oncorhynchus mykiss*, 63–65%) [30], Atlantic salmon (*Salmo salar*, 62–69%) [31; 32], river catfish (*Pangasianodon hypophthalmus*, 34%) [33] and sea bass (*Dicentrarchus labrax*, 34%) [34]. Tilapia has advantage over other aquatic species in the number of the breeding programs around the world [26; 28; 21; 29; 27; 35]. (The known number of the tilapia breeding programs in operation is at least 20 families based breeding programs [28; 21; 35; 27]. Stronger competition with other tilapia producers increases the necessity to reduce production cost and improve quality of tilapia products. In addition to improving the production systems, it is also necessary to develop genetically improved breeds of tilapia that perform well under different production systems. The ultimate objective in any commercial breeding program is to improve the biological efficiency of an animal production by utilizing the genetic variation among individuals. The breeding objectives, which define the direction of genetic change in the breeding population, should ideally include all traits that are commercially important in commercial tilapia farming. It is also necessary that these traits are measurable with available technologies, that the recordings are not too expensive and that the traits show additive genetic variation. The breeding programs for Nile tilapia will have breeding objectives that include some common traits and some specialized traits. Traits that should be discussed as breeding objective of Nile tilapia, some traits (i.e., feed efficiency and robustness) are not included in the selection index, but will be indirectly improved by selection on correlated traits. Examples of the breeding program with specific objectives are:

- Feed efficiency
- Growth rate
- Robustness
- Salinity tolerance
- Temperature tolerance
- Disease resistance
- Fillet yield
- Fillet color
- Slaughter yield
- Age of sexual maturation

Recently, prodcucers are interested in fulfiling the consumer preferences, so they start to adopt breeding programs for quality tarits, such as fillet yield, slaugter yield and fillet color.

Fillet Yield

Tilapias are sold either as whole fresh fish, portion-sized fillets (31-33%) or as frozen blocks of white fillets (25-27%). The profitability of producing fillets will depend on size of fillets and fillets yield (i.e., fillet as a percentage of whole body weight). The size of fillets depends strongly on the slaughter weight of tilapia and will, therefore, indirectly be improved by a selection to increase growth rate. Fillet yield will, however, depend on body shape of tilapia – i.e., head size, body width, body depth, etc.

- The breeding goal is better fillet yield for Nile tilapia

Slaughter Yield

If tilapias are sold fresh and gutted profitability will depend on size and the slaughter yield (i.e., weight of gutted fish as a percentage of whole body weight).

- The breeding goal is better slaughter yield for Nile tilapia.

Fillet Color

There is different color morphs of tilapia and the fillet colors depends on the color morphs. There is red strain, silver strain, and molted tilapia strain. Color can be produced by hybridiation between different color morphs and then used in the breeding program for selection of fillet colr and thickness.

- The breeding goal is different fillet colors for Nile tilapia

Commercial Breeding of Fillet Yield for the Market Needs

Frozen tilapia filets, mainly derived from Asia and Latin-America, and exported to USA, Europe and other countries, are now a world-wide commodity. With the developing export markets for frozen filets during the 1990s, the desired harvest weights increased to 600–800 g. Fillet yield is considered an important trait affecting the economic efficiency of production systems in many commercial production enterprises of Nile tilapia [3; 29; 36]. However, fillet and carcass yields have great economic importance for the meat processing industry. There is little published information regarding genetic variance components for fillet yield of Nile tilapia. There is a need for developing genetic studies to develop fillet yield in Nile tilapia, this is because of low [3] to moderate [29] heritability estimates of fillet yield. Knowing of genetic and environmental covariance structures during the growing period could help to establish the best time to measure and select fish to obtain the maximum rate of selection response for fillet yield in Nile tilapia. It also would be useful to predict correlated response in each trait to direct selection for any other, body weight, e.g, that would not require slaughtering to allow data collection and could result in increased accuracy of predicted breeding values for the analyzed traits [37]. The estimated genetic correlations among body weight, fillet weight and carcass weight exceeded 0.98 and were similar to estimates observed in the literature [29; 3; 38]. Results from the literature that use body measurements as a selection criterion are not conclusive but suggest no improvement of fillet yield but increased fillet weight in Nile tilapia populations [29; 38; 3; 39]. In commercial tilapia breeding program, [36] states that considering fillet weight as selection criterion and target trait of a breeding program, selection should be performed at around 200 days post-hatching; because the genetic gains will be higher and the genetic correlations are higher than 0.97, thereby avoiding slaughter of fish for data collection purposes and ensuring higher selection pressure. The economic value of improving fillet yield is of considerable importance for the processing industry that purchases tilapia as whole fish. Fillet yield of Nile tilapia has become an even more important trait as processors implement additional trims and deeper skinning at the request of some costumers [7].

Selection Methodologies

In tilapia breeding programs for market preference, the important breeding objective traits are fillet thickness and fillet color. Heritability of low (0.12 ± 0.06, [2; 3] to medium (0.25 ± 0.07; [29] magnitude has been reported for fillet yield in Nile tilapia, indicating that the fillet yield can be improved through selection for increased growth. Due to high genetic correlations (close to unity) between body weight and fillet weight [2; 3; 29], implying that selection for body weight will improve the fillet yield [41]. This recommends for establishing breeding programs with one goal (fillet yield/ or body weight) [40]. Selection for fillet yield requires required fillet trait (fillet weight and color) on live breeding candidates, so selection for fillet trait will be on their sibs. Selection for fillet trait could be recorded on correlated traits (such as boy weight) and body shape [42; 43; 44]. However, if we can be able to record the fillet traits on live animals, the selection intensity and accuracy will be increased which will in turn increase the genetic gain as compared to sib selection. In the same time exploits the genetic variability between and within families. The success of the tilapia breeding programs will depend on establishing base populations of a general high genetic quality and with a brood additive genetic variation for all traits included in the breeding objectives. Selective breeding programmes are only appropriate when genetic variation exists for the traits to be selected. A number of breeding programs in fish may have failed due to low genetic variation in the base population [45;

46]. A list of available genetic material should be set up before the base population is formed. The additive genetic variation of the base population of Nile tilapia should be secured by collecting genetic materials from different sources. It should be an open breeding nucleus where new genetic material can be introduced at any time. New genetic material should be tested as families in the breeding nucleus. After testing and evaluation of "new" families strategic decision about "new" genetic material should be taken. It is essential with a protocol to avoid import of diseases.

Different strategies of selective breeding use different sources of information to estimate breeding values of the breeding candidates. A breeding value is defined as the ability of an individual to produce high or low-performing offspring. The choice of strategy to be used in the present programs is decided based on the biology of the tilapias and the characteristics of traits that were included in the breeding objective. Selection based on individual information, which often is referred to as "mass –selection" is only using information observed on the breeding candidates themselves to estimate their breeding values. It follows that this strategy can only applied on traits that are recorded on live individuals (i.e., the breeding candidates), and it is only efficient to improve those traits that show a medium to high heritability (i.e., when relatively much of the observed variation between the breeding candidates is due to differences in their breeding values). Although selection based on individual information can be effective to improve some traits, such as growth rate (recorded as boy weight at harvest), the long term selection response is often reduced due a too fast accumulation of inbreeding in the breeding population. Such accumulation of inbreeding might both reduce the performance of fish due to inbreeding depression and reduce the potential of further improvement due to reduced genetic variation. Therefore, individual selection requires measures to restrict the accumulation of inbreeding in the breeding population.

In family selection, family members (i.e., full and half-sibs) can also be used as sources of information to estimate breeding values of the breeding candidates since they carry, in average, 50% (full-sibs) and 25% (half-sibs) of the same additive genetic material. This selection strategy usually referred to as family selection and is especially efficient to improve traits with a low heritability (i.e., a small proportion of the observed variation in a trait is due to additive genetic variance) or show a binary distribution (e.g. survival rate). Since it is possible to produce large families of Nile tilapia, the breeding values based on family-information can have a high accuracy. However, family selection requires separate rearing of families and individual tagging to keep track of their pedigree. When traits cannot be recorded on live fish or traits with low heritability is included as breeding goal in the breeding program family selection is the only alternative for these traits. Combined selection can applied when several traits are included in the breeding objective, as in the present breeding programs on tilapias, a selection based on both individual (i.e., the breeding candidates) and family (i.e., full- and half-sibs) information should be used to maximize the selection response. Combining different information sources will allow a simultaneous selection for all traits that can be measured, including those traits that cannot be recorded directly on the breeding candidates themselves (e.g. fillet yield and growth performances in different test environments). This selection strategy will also secure a highest possible accuracy of the estimated breeding values and maximize the selection response. Therefore, all data from the breeding candidates and their relatives will be collected and statistically analyzed as a genetic trend analyze in the present breeding programs. Multi-trait improvement is used for several traits economically important in tilapia farming and should, therefore, be included in the breeding objective. Since a simultaneous selection to improve several traits is more efficient than to select individual traits one after another (i.e., tandem selection). The selection of breeding candidates should be conducted based on a selection index of the traits in the breeding objectives. Each trait should be weighted according to its economic value in the selection index to assure that the tilapias are selected based on their overall economic breeding value. A selection based on such index values will make it possible to select breeding candidates that will produce offspring of a higher economic value.

Selection Strategies

[2; 3] reported considerably higher fillet yields in GIFT tilapia compared with other breeds of Nile tilapia (i.e., 38.0%, based on fillets without scales, but with skin and ribs), while [30] reported lower estimates of fillet yield (i.e., 31.1–34.9%, based on manually skinned fillets) in different generations of GIFT tilapia. While [29] observed no significant effect of age and [2; 3] reported a positive, linear effect of age on fillet yield, estimates of fillet yield in the present breeding program were influenced by both positive linear and negative quadratic effects of age at recording when analyzing all data. Interestingly, recorded fillet weights in a study by [36] were positively influenced by body weight and negatively influenced by age at recording.

Predicted fillet yields at different ages (6 and 9 months) suggest that younger individuals have higher fillet yield than older individuals, and that the positive effect of body weight is more significant in older individuals. This could explain the results reported by [39], which showed considerable higher fillet yield in the younger and larger-sized base population (G0: 45.1% fillet yield, 1200 g body weight, 202 days age) than in the following selected generations (G1: 41.3%, 650 g, 276 days; G2: 42.2%, 860 g, 250 days), and also the large linear effect of body weight (i.e., 0.1%-points higher fillet yield per 100 g increase of body weight) reported for very old Nile tilapia (426 days) by [44]. This suggests that not only body weight at filleting is important for fillet yield, as has also been reported for other aquaculture species [47], but also age at filleting is of major importance. The negative effect of age on fillet yield could be due to increased visceral percentage [30] or other physiological changes caused by sexual maturation and breeding activities. In general, the estimates of fillet yield increased with increasing number of selected generations in the present breeding program. The much lower mean estimate of fillet yield in the conducting the filleting operation, as also reported by [29]. Breeding programs for Nile tilapia have mainly focused on improving growth by mass-selection [17], within-family selection [48] and by combined family and within-family selection strategies utilizing full pedigree information[1; 49; 4; 40]. Mass-selection and within-family selection can only be used to improve traits that are recorded on the breeding candidates while they are alive (i.e., body weight, body shape, fin stripe pattern, etc.…) and hence can be used to produce a new generation of offspring after the recording is made. Since fillet traits can only be recorded on sacrificed fish (typically close siblings of the breeding candidates), several studies have investigated whether it is possible to improve fillet yield through indirect selection for body traits recorded on live breeding candidates [29; 44; 2; 3].

Unfortunately, improvement of fillet yield through indirect selection for body traits seems to be very ineffective across species (reviewed by [29], and it has been argued that it is better to focus on improving the fillet weight through indirect selection for increased body weight. Direct selection to improve fillet weight and fillet yield requires a large breeding scheme where the two traits are being simultaneously improved using a combined family and within-family selection.

Heritability and Genetic Correlations between Fillet Traits

The heritability estimates of skin-on fillet weight were low to moderate in magnitude and highly variable among generations (0.00–0.45) partly because of apparent confounding between additive genetic and common environmental (c^2) effects as was also observed for body weight at harvest [40]. Such confounding between additive and c^2 effects is probably caused by restricted pedigree information or insufficient genetic linkages between sub-classes of data. The overall heritability estimate of skin-on fillet weight (0.30) was comparable with or larger than those reported for Nile tilapia [39; 29; 2; 3] and rainbow trout [30], and higher than the earlier reported h^2 estimate for body weight at harvest of the same Nile tilapia population [40]. Full-sib effects other than additive genetic effects (c^2) of fillet weight (0.08) and body weight at harvest (0.10) were of similar magnitude. The heritability estimates for the different estimates of fillet yield (i.e., based on skin-on fillets, skinned fillets and trimmed fillets) were low to moderate in magnitude (0.08–0.30) and comparable with estimates reported in literature [39; 30; 21; 29; 32; 2; 3]. As also reported in most of the above cited literature, c^2 of different estimates of fillet yield were very low in magnitude (0.01–0.03) when analyzing all data. The genetic correlations between different estimates of fillet yield were all very high (0.95– 0.97 when analyzing all data) confirming that they are all reflecting the same underlying trait irrespective of processing level. Since increased processing levels (i.e., skinning and trimming) had no significant effect on the heritability estimates, it was decided to use fillet yield based on skin-on fillets as the selection criteria to make recording simpler and less resource demanding.

Genetic Correlations between Fillet Traits and Growth

The genetic correlation between skin-on fillet weight and body weight at harvest was, generally, very high in magnitude as has also been reported for Nile tilapia in literature [39; 29; 2; 3]. Selection for increased body weight at harvest will, therefore, efficiently improve fillet weight of Nile tilapia. However, the efficiency of producing tilapia fillets will also depend on the genetic correlation between fillet yield and body weight at harvest. Earlier studies have reported favorable genetic correlations between fillet yield and body weight at harvest of medium (0.44 ± 0.20; [29] or high magnitudes (0.74 ± 0.18; [2; 3]. Estimates reported in the present breeding program were not favorable, however, with a non-significant genetic correlation between fillet yield and body weight at harvest (0.09 ± 0.14) when analyzing all data combined. The genetic correlation between fillet yield and body weight at harvest shows large variation across breeding populations and generations within the same breeding population probably due to recording a limited number of test fish from different grow-out environments, of different sex, age and body weight at recording, and different skills of filleting technicians [30] or accuracy of filleting machines [2; 3]. It is also possible that the very high genetic correlation between fillet weight and body weight would restrict any systematic variation of fillet yield compared with the random variation caused by the factors listed above. Furthermore, the much lower phenotypic variation of fillet yield than that of body weight could cause even more uncertainty to the estimated genetic correlation between these traits. Finally, it is also possible that the genetic correlation between these traits may change with the selection to improve growth rate. In the present study, the relationship between estimated breeding values for fillet yield and body weight was non-linear and variable from one generation to another. [51], who reported a negative genetic correlation between fillet yield and body weight of gilthead seabream (−0.59), recommended that fillet yield should be monitored in fish selected for faster growth.

Selection Responses

Generally, the selection response for a given trait is determined by the selection intensity, the h^2 and the phenotypic standard deviation of the trait [41]. Indirect selection responses in other traits are also influenced by the h^2 and the phenotypic standard deviation of these traits and their genetic correlation with the first trait. The estimated indirect selection response in skin-on fillet weight (121 gram) was considerable higher than the expected gain based on the accumulated selection response in body weight at harvest[40] multiplied with the LS mean skin-on fillet yield across generations (i.e., 205 g × 44.4% = 91 g). This suggests that, although the genetic correlation between fillet weight and body weight at harvest was very high, the combined selection to improve both body weight at harvest and fillet yield has utilized the broader genetic variation (as observed by a higher h^2 and larger CV) of fillet weight compared with that of body weight at harvest to increase the fillet proportion more than the whole body weight. [29] reported that selection for high growth (recorded at about 250 g mean body weight) had no effect on the fillet yield of Nile tilapia, despite a positive genetic correlation, possibly due to physiological limitations. However, results in the present breeding program suggest that the fillet yield of Nile tilapia can be genetically improved when including fillet yield together with growth (recorded as body weight at harvest) in a multi-trait selection. The low selection response of skin-on fillet yield (0.2%-units per generation) was comparable with that reported for carcass yield (0.3%-units per year) and breast yield (0.2%-units per year) of chicken [52]. Although these estimates suggest a very slow genetic change of body traits per generation of selection, the accumulated gain may become significant in the longer term as has been observed in modern farmed chicken. Furthermore, the estimated selection response of fillet yield (1.2%-units) was too low to explain the estimated selection response of fillet weight. Based on the selection response estimates of fillet weight (present study) and body weight at harvest [40], the accumulated selection response of fillet yield could be considerable larger (5.4%-units).

Implications for Further Selection

In general, a selective breeding program should focus on genetically improving economically important traits with a long-term perspective [53]. However, efficient programs should also make necessary changes in the breeding objective as new information about the traits and production conditions are obtained. The present breeding program was established to meet the interests of both tilapia producers and processors. Given the available information at that time, it was decided to initiate a multi-trait selection to improve both growth and fillet yield without considering their relative importance to improve the efficiency of fillet production. Based on the selection responses achieved for growth [40], fillet weight and fillet yield, it can be argued that the improvement of fillet yield had marginal importance compared with that of growth to improve the efficiency of fillet production. Although fillet yield was genetically neutral to growth, the multi-trait selection restricted the selection response of growth and possible correlated effects to reduce production costs [54]. Furthermore, as long as tilapia is sold based on their whole body weight, the farmers have no economic incitement and hence are not willing to buy more expensive seed to farm tilapia with higher fillet yield. Finally, the increasing problems with unstable farming conditions and disease outbreaks cause tilapia farmers to focus more on producing tilapia with higher survival rate rather than higher fillet yield. Therefore, the future breeding objective should substitute fillet yield with traits (i.e., disease resistance) that improves the robustness of Nile tilapia.

Genetic parameters for body size measurements can be used to predict correlated responses in fillet- weight and fillet-yield [29; 44]. A few studies in Nile tilapia have reported genetic parameters for body length, height and thickness. In general, genetic correlations of these traits with harvest weight and fillet weight are high while genetic correlations with yield are much lower [55; 39; 2; 3]. Fish body shape is becoming of increasing interest to consumers and producers [56; 57]. Consumers are willing to pay higher prices for well-shaped fish, which is especially true for live fish and un-gutted fish. Condition factor, defined asweight/length3, is the most common parameter used to express shape in fish. Condition factor has been studied in many species, e.g. rainbow trout (*Salmo gairdneri*) [58], gilthead seabream (*Sparus auratus*) [51] and olive flounder (*Paralichthys olivaceus*) [59], but not in tilapias. However, condition factor describes only the relationship of the weight with a given length; it does not clearly describe the appearance of fish. To better describe shape, appearance, described as ‘slender’, ‘medium’ and ‘rotund’, has been proposed [57]. Heritability estimate of this shape trait in rainbow trout was 0.33 [57]. More recently, ellipticity was used to describe shape in common sole (*Solea solea*). [56] reported a heritability of 0.34 and showed that selection for harvest weight would lead to a undesired correlated response in shape. Genotype by environment interaction (G×E) has been studied extensively in Nile tilapia [20; 60; 61; 40]. Most G×E studies with Nile tilapia found no evidence of biologically important G×E for body weight as indicated by genetic correlations ranging from 0.73 to 0.99. Genotype by environment interaction has also been studied for other important cultured species, e.g. common carp (*Cyprinus carpio*) [62], rainbow trout [63; 64], and other salmonids [65]. Interestingly, in European seabass, substantial G×E was found for growth rate while no G×E was found for body weight at harvest [66].

Protection against “Genetic Linkage” and Unauthorized Reproduction

It is difficult to avoid that other tilapia hatcheries are stealing genetic material from the tilapia breeding populations. Tilapia breed easily and even commercial “all – male” tilapia can be used by competing hatcheries to improve the genetic quality of other broodstock. An efficient organization of dissemination that both secures a highest possible genetic quality of the commercial seed and reduces the risk of losing genetic material to competing tilapia hatcheries should be set up. This is achieved by producing two or more lines of commercial breeders based on genetic material from only few families in the breeding programs. The commercial seed will be produced using males and females from different lines to avoid any negative effects cause by inbreeding. The commercial seed will, however, represent a very limited genetic variation and soon cause severe inbreeding if these are used as broodstock in competing hatcheries.

Practicalities and Protocols of the Selective Breeding Program

Production of Families

Since many important traits (e.g. disease resistance, fillet yield and growth performance in alternative production environments) cannot be recorded on the breeding candidates themselves, it is necessary to produce families in separate rearing units from which a representative sample of breeding candidates and test fish should be tagged and tested for different traits. Test fish are mainly used as a source of information when estimating breeding values of the breeding candidates. The families have to be kept in separate hapas/tanks from first feeding until tagging.

Families Produced in Batches (Cohorts)

Batches of full-sib and half-sib families are produced continuously to allow adequate selection across batches. The families within each batch have to be kept in separate hapas/tanks from first feeding until tagging. Production of families from spawning to tagging will take 4-5 months and genetic links between batches have to be set up.

Families Produced Once a Year

For this alternative, full-sib and half-sib families are produced once a year. The families within each year have to be kept in separate hapas/tanks from first feeding until tagging. Production of families from spawning to tagging will take 4-5 months. Producing 200 families within the time period (1 month) necessary for a good comparison of families will most likely require partition of families into minimum two batches.

CONDITIONING BREEDING

After harvesting and recording, the pre-selected breeders should be sexed and male and female tilapia transferred to separate large hapa-cages for conditioning. This procedure is necessary to avoid uncontrolled breeding and ease the final selection and handling of selected breeders during the breeding period. Tilapia should be fed a balanced high quality tilapia feed.

SYNCHRONIZING THE SPAWNING

It is important to synchronize the spawning of females and, thereby, the production of families to reduce the confounding effects of age when estimating genetic parameters and breeding values. The spawning of tilapia can be synchronized by conditioning the potential breeders in the same environment (i.e., large hapas-cages), giving an environmental trigger – signal by changing the water temperature (first increasing the water temperature by reducing the water level in the pond, and then decreasing the water temperature by adding water) and feeding level (reduce daily feeding to 0.6% of body weight and then increase it to 1.2% of body weight during the following week), evaluating the sexual maturity of female breeders (as described in Figure 8) and stocking selected male breeders first together with selected female breeders that are in the " Ready to spawn " stage (i.e., RS families).

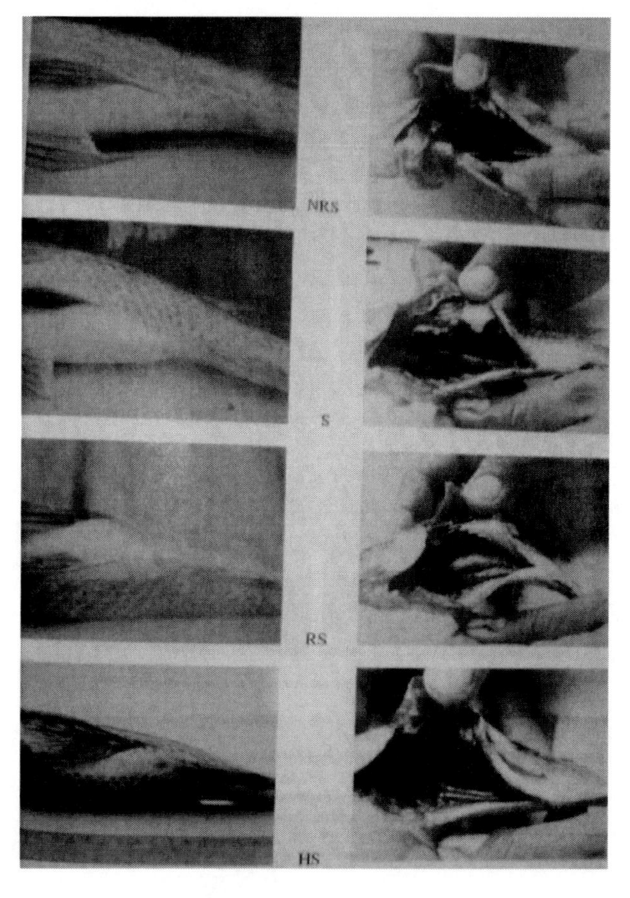

Figure 8. Evaluation of the sexual maturation of female tilapias using the categories given in Table 1: "NRS" Not Ready to Spawn, "S" Swollen, "RS" Ready to Spawn, and "HS" Has Spawned.

NATURAL BREEDING

Full –sib families should be produced by natural single-pair mating of selected male and female breeders in separate tanks. The hapas/tanks should be covered with shading net to secure an optimal and stable water temperature for breeding. Selected male breeder should be used to fertilize eggs from two selected female breeders to produce the families. The following procedure should be used: first conditioned RS- females (Figure 8) are transferred to separate breeding hapas/ tanks, i.e., one female breeder per breeding hapa/tank, and then a conditioned male breeder is stocked in each of the occupied hapa/ breeding tanks. After mating, the male breeders are transferred to another RS-female to produce their second full-sib family. All the breeding hapas/tanks should be inspected regularly, especially at the expected time of spawning, to find out whether the female breeders have spawned. The female breeders will usually incubate spawned eggs and newly hatched fry in their mouth. Large female breeders might, however, swallow their eggs if they are being too much stressed by the male breeder. Therefore, fertilized eggs or plum-sac larvae should be collected from each breeding tank and be separately incubated in small incubators/trays; one unit per full-sib family (Figure 9). Swim- up fry should be transferred at a standardized number to nursing hapas/tanks; one nursery hapa/tank per full-sib family.

Figure 9. Tilapia eggs and larvae of different full-sib families should be incubated in separate incubators/trays. Swim-up fry should be transferred to nursing hapas.

Salinity Tolerance

Test for salinity tolerance might be growth and survival in brackish water, or in controlled challenge experiment. I developed protocol for a temperature challenge test used for red tilapia in Egypt (Figure 10), which proved operational and produced nice results.

Figure 10. Challenge tests for Salinity and temperature tolerance of Nile tilapia.

Temperature Tolerance

A protocol for controlled experiments aimed at recording of tolerance to low temperature need to be developed. Costs expected to be reasonable.

Disease Resistance

Likewise, detailed protocols and facilities need to be developed for controlled challenge tests for recording genetic variability in resistance to key pathogens. Routine challenge tests against specific diseases are currently implemented for salmonids (Norway), marine shrimp (Colombia/USA), carp (India), tilapia (Thailand).

Reducing Aggressive Behavior of Male Breeders

Nile tilapia males usually have a very aggressive behavior and large individuals might even kill the female breeders in connection with single-pair mating in small breeding hapas/tanks. To reduce their aggressive behavior, it might be necessary to carry out mouth clipping (Figure 11) and, if this is not enough, also fin clipping of male breeders before transferring them to the breeding tanks. The male breeders should be anesthetized before removing their upper lips (incl. the bone structure) and fins, and the wounds should be disinfected using an antiseptic preparation.

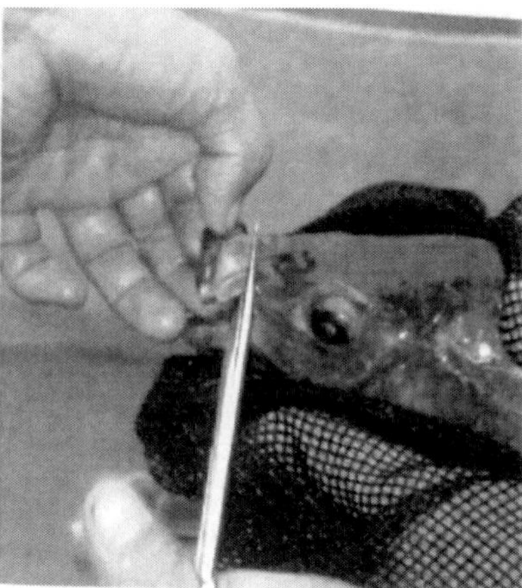

Figure 11. The male breeders of Nile tilapia should be mouth, and possibly, fin clipped to reduce their aggressive breeding behavior.

REARING OF FAMILIES

Each full-sib family of tilapia should be reared in separate hapas/tanks until they large enough to be tagged (12-15 g). A random sample of swim-up fry from each family should first be transferred to separate nursing tanks/hapas. After 21 days of rearing, the fry are transferred to B-net cages (larger mesh nets) that will allow better water circulation and improved growth, and reared until fry of the youngest family reach an average body weight of 8-10 grams, at which body size they can be individually tagged. The fry/fingerlings should be fed a formulated floating feed according to their appetite.

Individual Tagging

Fish from all families should be tested in the same production environments to reduce environmental effects and, thereby, increase the accuracy of the estimated breeding values. The maximum time for production of a batch should be 4 weeks and the production of families within a batch should be terminated after 4 weeks of independent of number of families produced. To keep track of the pedigree of all breeding candidates and test fish, it is necessary to individually tag the tilapia as soon as they become large enough to be tagged. The present breeding programs use Passive Integrated Transponder (PIT) tags to individually tag breeding candidates and test fish of tilapia (Figure 12), as used in many other fish breeding programs worldwide. The PIT tag consists of an antenna coil encapsulated by a glass ampoule. The reading device used to record the PIT tags creates an electromagnetic field that energizes the tag within its read range and makes the transponder return a unique ID via signal modulation. The reader in turn amplifies this ID code, converts it to digital form and decodes it as an alphanumeric number that is stored electronically on a computer.

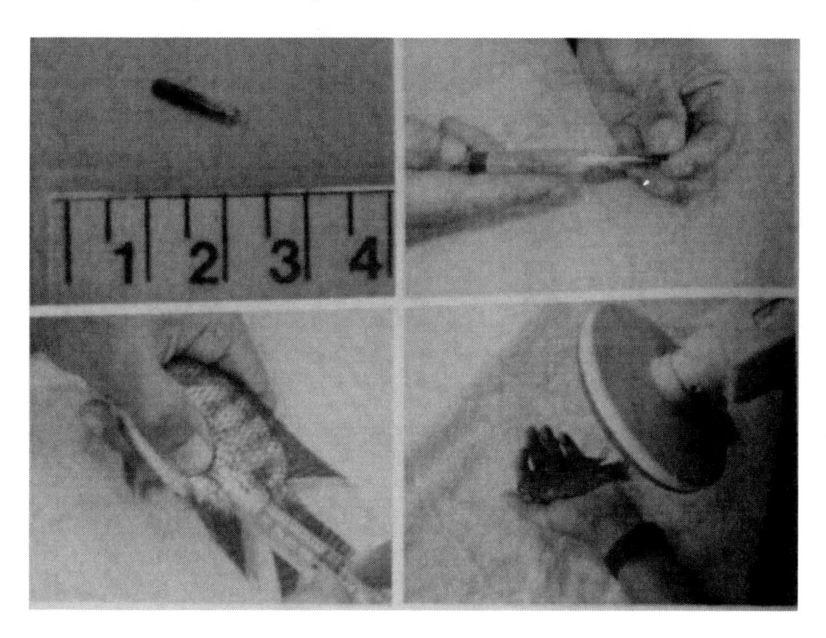

Figure 12. Test fish and breeding candidates should be individually tagged using Passive Integrated transponder (PIT) tags.

A number of randomly selected individuals from each full-sib family should be PIT-tagged as soon as the youngest families reach an average body weight of 12-15 grams. During the tagging operation, the fish are anesthetized and the tag is implemented in the abdominal cavity of the fingerling intra-peritoneally through a needle attached to a spring-loaded syringe. During the recovery phase after tagging, the fish should be kept in aerated water and if necessary treated to prevent infection. All fish should be kept under observation for at least 24 hours before transfer to the test environments. Mortalities and the frequency of tag rejection in connection with the tagging operation are expected to be very low. Tags should be recovered from harvested fish and reused to tag other batches.

TESTING

After tagging, representative samples of fish from all full-sib families should be tested for the different traits (or correlated traits) as defined in the breeding goals for Nile tilapia. The following describes the procedures that should be followed when testing different traits.

Performance Test Ponds/Cages

Test –fish pond and test-fish cage will be tested for growth performance (i.e., growth and survival) slaughter yield, filet yield and quality. The test fish will be harvest as close to market size as possible, however the test fish have to be recorded and the results processed before selection of the breeding candidates.

BREEDING CANDIDATES

When results from test-fish are ready the breeding candidates in pond/Cage will be recorded and preselected. The breeding candidates will be final selected and sexed before transfer to conditioning hapas (Figure 13).

Figure 13. Hapas used in conditioning the breeding candidates.

Recordings

Efficient selective breeding programs will require accurate recordings during production of families, and at tagging and testing of breeding candidates and test fish. Therefore, it is very important to develop good routines for making such accurate recordings. The following is an overview of information that should be recorded in the present breeding program.

At Breeding

Families (i.e., full- and half-sib family groups) will be produced in breeding tanks/ hapas where one male breeder will be stocked together with one female breeder for single – pair mating. The following should be recorded in connection with this natural mating:

- Tilapia species (Nile tilapia)
- Batch number or year class
- Breeding tank/hapa (family) number
- Individual ID –number (PIT-tag) of female breeder (dam)
- Individual ID –number (PIT-tag) of female breeder (sire)
- Date at stocking of dam and sire in breeding tank/hapa
- Date at collection of eggs/yolk-sac larvae

During Family Rearing

The collected eggs/yolk- sac larvae from each breeding tank/hapa (female) will be incubated and the swim-up fry transferred to at a standardized number of individuals per family to nursing tank/hapas, where they will be reared for three weeks. Fingerlings will be transferred from the nursing hapas at a standardized number fry per family to B-net cages and reared until the youngest family reach a suitable size for tagging (i.e., an average body weight of 12-15 grams). The following should be recorded during this period:

- Tilapia species
- Batch number or year class
- Breeding tank (family number)
- Date at collection of eggs/yolk-sac larvae
- Egg incubator/tray number
- Date at transfer of swim-up fry to nursing hapas
- Nursing hapa/tank number
- Survival in each nursing hapa
- Date at transfer of fry from nursing hapas to B-net cages
- B-net cage number
- Survival in each B-net cage

Tagging

When the youngest family reaches an average body weight of about 12-15 grams, random samples of fish from all families produced in the batch should be PIT – tagged. The following should be recorded at tagging:

- Tilapia species
- Batch number or year class
- Date at tagging
- Breeding tank (family) number
- Individual ID- number (PIT – tag)
- Individual body weight
- Test environment or breeding candidate
- Stage of sexual maturation

Recording of Growth Performance

- Tilapia species
- Batch number or year class
- Test environment or breeding candidate.
- Date at stocking
- Water temperature (daily recordings)
- Feeding (daily recordings)
- Date at harvest
- Individual ID-number (PIT-tag),
- Body weight
- Sex

Recording of Quality

The test fish have to be sacrificed for the recording of fillet yield. The following should record in connection with recording of fillet yield:

- Tilapia species
- Batch number or year class
- Test environment
- Date at recording

- Individual ID- number (PIT) tag
- Body weight
- Sex
- Size of gonad
- Gutted weight
- Fillet yield (skin on)

Pre-Selection of Breeding Candidates

At harvest, breeding candidates from the 50% highest – ranked families will be pre-selected (based on their families breeding values), sorted by sex and transferred to large hapas/cages for preselected breeding candidates. The following should be registered in connection with pre-selection of the breeding candidates:

- Tilapia species
- Batch number or year class
- Test environment
- Date at harvest and pre-selection
- Individual ID-number (PIT-tag)
- Individual body weight and sex at harvest
- Destiny (i.e., pre-selected or not)
- Hapa /cage number

Final Selection of New Breeders

Breeders to produce the next generation of families in the tilapia breeding program will be selected based on their individual economic breeding values including all recorded traits. After final selection the selected breeders will be transferred to conditioning ponds. The following will be recorded in connection with the final selection of breeders:

- Tilapia species
- Batch number or year class
- Individual ID-number (PIT-tag)
- Destiny (i.e., selected to produce next generation in breeding program, selected to produce lines or killed).

Selection of New Breeders for Nucleus

The genetic gain in a selective breeding program (i.e., selection response) will depend on the additive genetic variation of the traits (i.e., variation of breeding values) in the breeding populations, the accuracy of the estimated breeding values and the selection intensity.

Estimation of Breeding Values

Heritability estimates and breeding values of all recorded traits in the breeding objectives will be estimated using the " Best Linear Unbiased Prediction" (BLUP) procedure, which utilizes all available information by first calculating the additive genetic relationship between all breeding candidates and test fish. Combining several information sources will increase the accuracy of the estimated breeding values and maximize the genetic gain in the next generation of the tilapias. Individual breeding values shall be calculated for growth (recorded as body weight) of all tagged fish in (both test fish and breeding candidates), while family breeding values shall be recorded for the other recorded traits.

MULTI-TRAIT SELECTION INDEX

Estimated breeding values for all recorded traits will be combined in a multi-trait selection index to estimate a total (aggregated) economic breeding value of all breeding candidates. Different traits should be weighted in the selection index according to their economic breeding value in commercial tilapia farming. This will allow a selection of breeding candidates that produce offspring of a higher economic value.

Pre-Selection

The breeding candidates should be pre- selected based on preliminary selection index values calculated for each family (i.e., family breeding values) as soon as information from the test fish have been recorded and statistically analyzed. The pre-selection will be conducted in connection with recording of individual body weights of the breeding candidates. All pre-selected male and female tilapia will be restocked for conditioning in large hapa/cages to avoid uncontrolled breeding.

Final Selection

The final selection of the breeders should be based on their individual economic breeding values, which combine information from the breeding candidates and their relatives. Male and female breeders, which will be used to produce the next generation of families in the breeding program, should be selected among the highest – ranking breeding candidates. However, it is necessary to make sure that the selected breeders are representing a minimum of full-sib families to reduce the accumulation of inbreeding and secure that a sufficient additive genetic variation is maintained in the breeding population for further genetic improvement in later generations. A summary of key operations required in the breeding is given in figure

CONTROL TO RESTRICT INBREEDING

Inbreeding will usually reduce the performance of tilapias and other aquaculture species, especially if the accumulation of inbreeding happens too fast. Such inbreeding depression affects most fitness traits (e.g. number and hatchability of eggs), number of crippled larvae, survival and growth of larvae and adult fish, and feed conversion efficiency. Inbreeding will also reduce the additive genetic variation in breeding populations and, thereby, reduce the potential for further genetic improvement. Therefore, the accumulation of inbreeding in the tilapia breeding populations must be restricted by ensuring that no closely related individuals are allowed to mate each other (short term prospective), and that a largest possible number of individuals are used as breeders in each generation (long term prospective). The number of selected breeders used in each generation will depend on the number of families that can be produced and the mating design.

GENETIC PARAMETERS WITHIN NUCLEUS

Breeding Nucleus

With an efficient organization of dissemination, centralized breeding programs (with its multiplier units) could easily furnish the customized products. Each new generation of families in the breeding programs should be produced by superior breeding candidates that have been selected based on their total (aggregated) breeding value, which combines breeding values for each trait in the breeding objective according to their economic importance to the tilapia farmers in Egypt (Figure 14). Some tilapia producers might, however, put different importance to the traits included in the breeding objective. These farmers could be offered "customized seed", i.e., commercial fry/fingerlings that are produced by breeding candidates selected according to a different combination of recorded traits than used in the breeding programs (demand for genetically improved tilapia in all major tilapia producing areas). The maximum delay from obtaining a genetic improvement in the breeding programs until the tilapia farmers can utilize it as limited to only one generation (about 6 months).

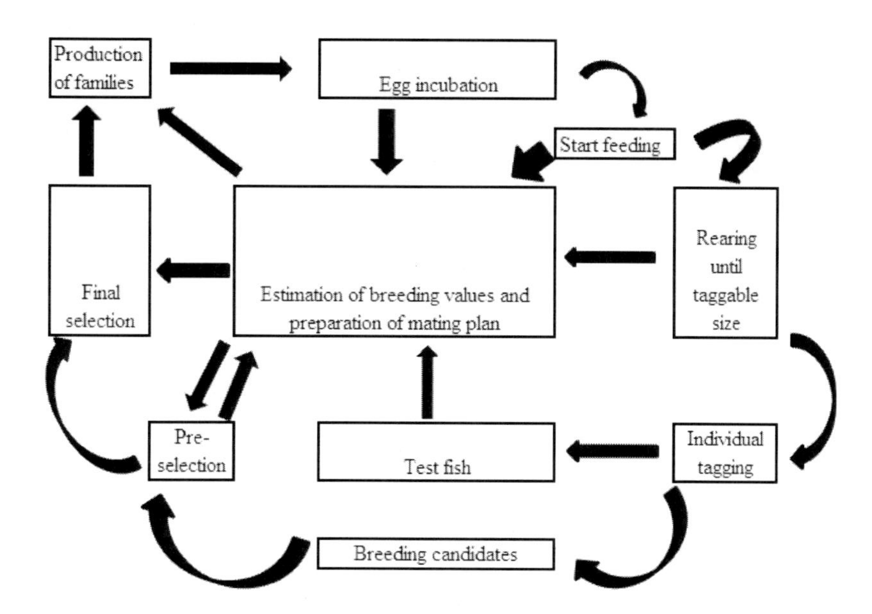

Figure 14. Summary of key operations required in the breeding nucleus for Nile tilapia (detailed mating plan to maximize selection responses and restrict inbreeding).

Applied Commercial Breeding Program for Quality Traits (Fillet Thickness and Fillet Color) in Nile Tilapia

Strain Comparisons and Selective Breeding

Summary

A commercial tilapia breeding program on was started in a private farm in Egypt in 2010 with the aim of improving fish growth, production and marketing value. The measured traits of the breeding program were fillet thickness and fillet color. The tilapia fish strains used in this breeding program were Taiwan Red, Stirling and EGYPT Silver. Selection was conducted on selection of fillet thickness. Thickness measured at the "shoulder" of the fish is a good indicator of fillet yield for a fish of a given size. Also, selection was conducted on fillet color and thickness on pure red tilapia and mottled tilapia fish. The breeding program work is designed as follow:

1. **Selection criteria:**
    1. Weight at age (growth rate);
    2. Width at age (fillet yield);
    3. Color;
2. **Random sample procedure for selection:**
    - (a) Randomly sample 100 fish;
    - (b) Measure weight (Wt) and width (Wd); record the data which is the baseline for selection;
    - (c) Grade the fish to produce 2000 individuals;
    - (d) Sample 130 fish from the 2000;
    - (e) Plot the Wt and Wd of the sampled fish;
    - (f) Do a best-fit regression line;
    - (g) Reduce the line by 10%;
    - (h) Select 650 fish of each sex;
    - (i) Reject 150 fish of each sex (30%) that least conforms to our color requirement.
3. **Strains:**
    - (a) Taiwan Red – clean up the Taiwan Reds. Propagate G0 to produce G1 and distribute the offspring in 2 places.
    - (b) Stirling – the G2 Stirling are contaminated, so clean them up by going through and selecting the "true-looking" Stirling. Mate the G0 Stirling and keep the offspring separate. Mate

the cleaned-up G1 and G2 and keep the *offspring* as G3. Distribute in 2 places.

4. EGYPT Silver – roll the ponds, breed, collect offspring and distribute in 2 places.
5. **Development of working broodstock:**

Strain	Generation
Egypt Red	G_0 G_1 G_2 G_3
Taiwan Red (pureline colour)	G_0
Stirling Red (pureline selection & wholefish market)	G_0 G_1 G_2
Egypt Silver (keep for reference)	G_2 G_0

6. **Batch Selection Template started** – a computer program to record and compare actual vs. projected performance in the selection program was started.
7. **Value of the program:** Approximately 2000 selected (G2) fish have been transferred into the production broodstock program. The fry produced by these fish (G3) will be compared with the EGYPT Silver reference. Therefore prepare a batch of silver fish to produce fry to put through the hatchery system starting in March.
8. **Resources required:**
   (a) A person (bright high-school?) for the Selection program and data entry for the Batch Selection Template;
   (b) A dedicated computer with CD burner and Windows XP;
   (c) An electronic scale (to 1 g) to weigh selected fish;
   (d) An electronic scale (to 0.1 g) to weigh small fish;
   (e) Millimetre callipers (2) to measure width of selected fish;
   (f) A fish measuring board;
   (g) Dedicated nets for bio-security purposes – 1 x 1/2" x 300' x 10' deep net for harvesting at 50 g, crowding, sampling and 1 x 1.25" x 300' x 10' deep square net for harvesting at the point of selection;
   (h) Nylon fishing line and side netting to bird-proof the ponds.
9. **Reporting.**

Final Outcome of the Strain Comparison Trials

The simplest way to compare the strains is simply to plot growth curves for all the data, as in Figure 15. Unfortunately this straight-forward procedure fails to account for environmental differences among the growth trials. Egypt silver fish were therefore grown together with the test strains as an internal reference to measure environmental quality. The test strains can be analysed relative to the standard strain growing in the same environment, whatever that environment may be.

Figure 16 shows the difference in mean weight of each strain and the reference strain, expressed in grams. Figure 16 differs from Figure 15 in that the data have been aggregated by week, that is, after reducing all the individual data to the mean differences between the weights of each strain and the silver strain, expressed as a percentage of the silver strain. It appears that although there is considerable scatter, the two "foreign" strains were larger than either the Egypt reds or the reference fish throughout most of their lives *when the data from all environments are considered together*. The Egypt reds were smaller than the contemporaneous silver reference fish until age 40 weeks or so, near the end of the experiment.

The relative weekly growth rates plotted in Figure 17 also show that *when the data from all environments are considered together* the Egypt red strain compares rather poorly to the other strains, even the Egypt silver reference strain, until late in the growth cycle. However, from an aquacultural point of view, it seems more appropriate to analyze the tank and pond environments separately, as shown in Figures 18 and 19 respectively.

In (Figure 1), There are too many data to be displayed (more than 5,000 points) so "lowess" smoothing curves have been drawn to represent the data on each strain.

In (Figure 16), data are aggregated by week. The horizontal dotted line represents the reference strain (no diference with itself).

In (Figure 18), data for the EGYPT reference strain are shown in black. A power curve has been fitted through the data.

In (Figure 19), data for the EGYPT reference strain are shown in black. A power curve has been fitted through the data.

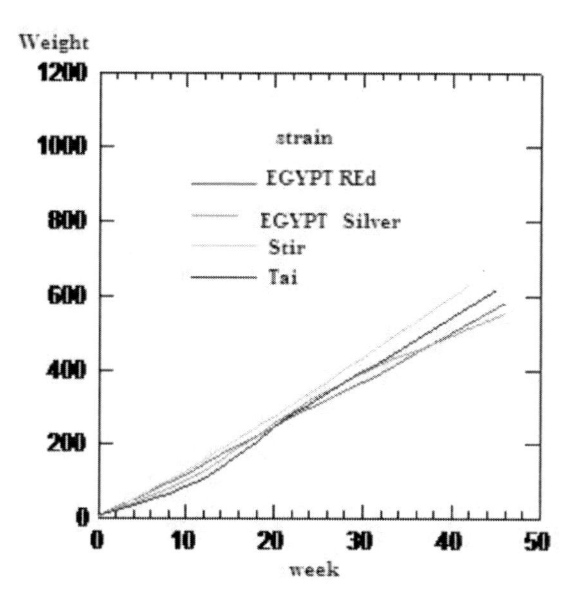

Figure 15. Growth curves for all strains in all environments combined.

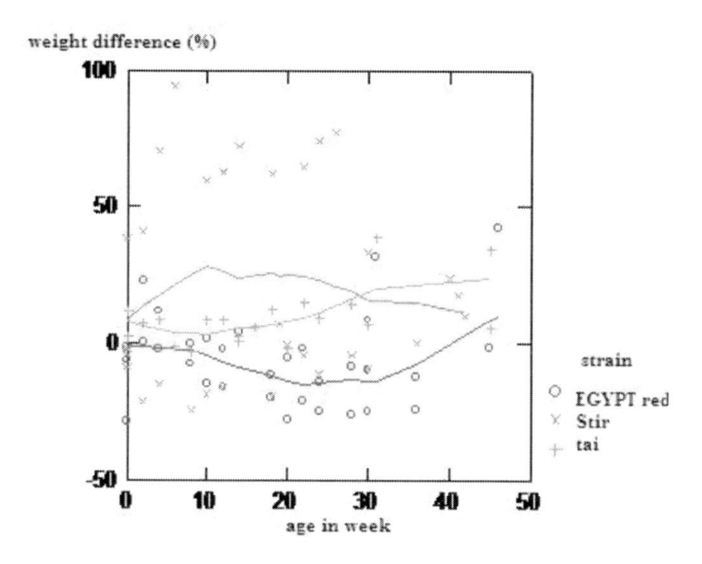

Figure 16. Percentage weight difference between test and reference strains in all environments combined.

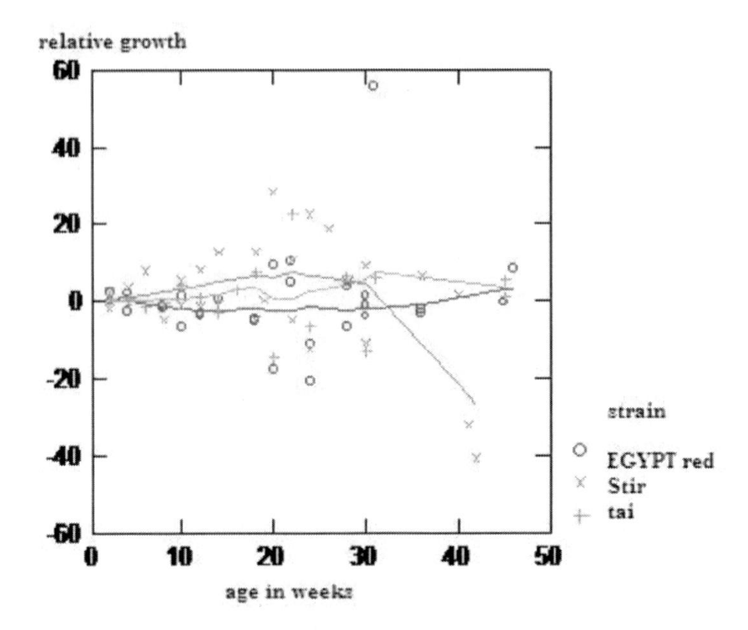

Figure 17. Growth rate difference between the test and reference strain in all environments combined.

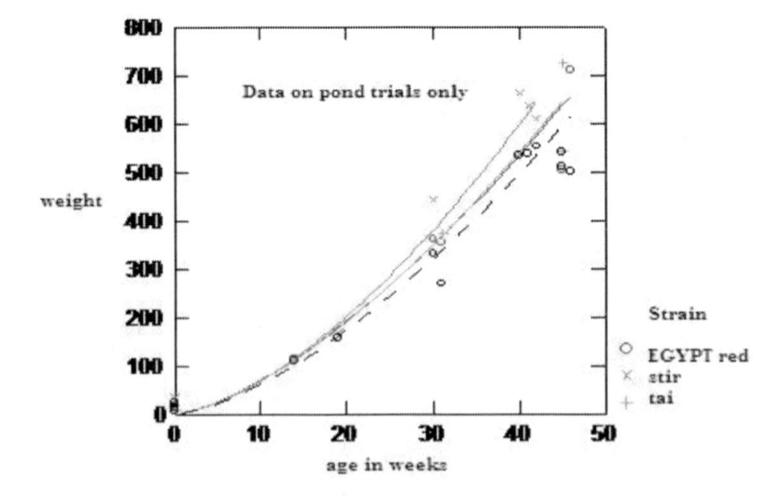

Figure 18. Growth curves for all strains in ponds only.

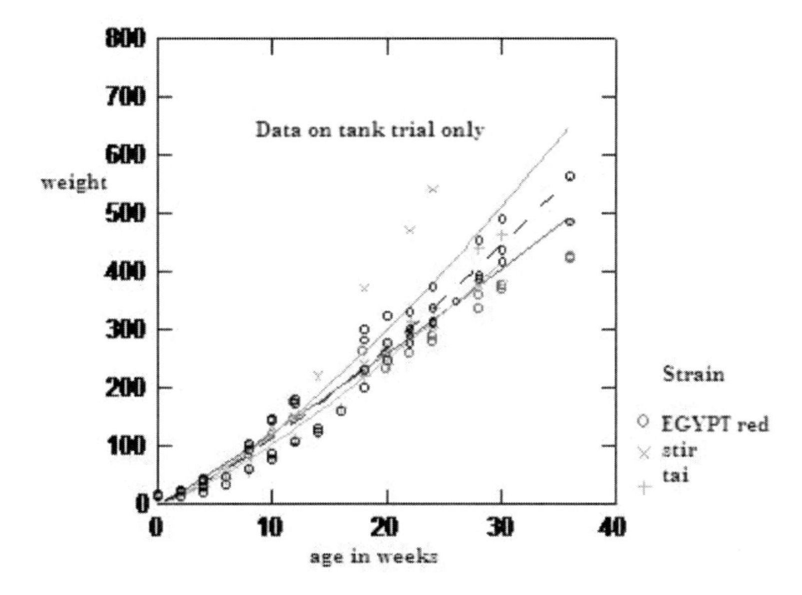

Figure 19. Growth curves for all strains in tanks only.

All the strains performed better than the reference strain in the pond environment (Figure 18). The Stirling strain performed best of all; at 40 weeks it was about 50 grams heavier, on average, than the EGYPT red strain. The Stirling strain was largest in the tank environment too (Figure 19), but it should be noted that all the Stirling tank data came from a single experiment. The other two test strains were smaller than the reference strains in tanks, for unknown reasons. Formally, this can be thought of as an example of genotype-environment interaction (G x E interaction) in which the rank of the strains changes in when they are grown in different environments. The growth rate of the two foreign strains, especially the Stirling strain, varied more than the EGYPT strains (Figure 20). Among the larger animals these strains achieved both the highest and lowest weight-specific growth rates relative to the reference strain. The variation was most pronounced when overall growth was slow, which suggests that the full genetic potential of the foreign strain may only be realized when conditions are suitable for rapid growth.

In (Figure 20), data for the EGYPT silver reference strain shown in black. *Specific growth* is the weekly gain in weight divided by mean body weight. Data include only those animals which weigh more than 100 grams.

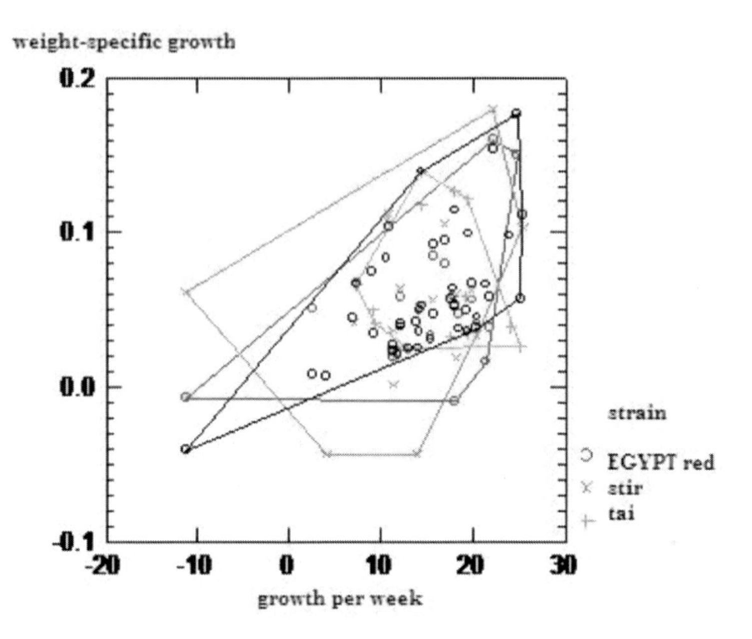

Figure 20. Variation in specific growth rate.

Note on Colour of Hybrids

[40] mentioned that tilapias are unable to synthesize carotenoids and rely solely on dietary supply to achieve their red/orange to yellow skin pigmentation. [67] stated that the external color of tilapia is primarily dependent upon the morphology, density and distribution of different chromatophores within the three dimensional organization of the skin. Five main types of chromatophore are commonly distinguished: melanophores (black–brown color caused by melanin), erythrophores (red–orange color caused by carotenoids), xanthophores (ocher–yellow color caused by pteridines), iridophores (metallic iridescent caused by purine crystals) and leucophores (white/creamy light reflection caused by poorly organized purine crystals). Studies carried out by [68; 69; 70] of color-mutated tilapia have all suggested that the body color is a simple trait governed by very few (1–3) genes. These genes have been reported to cause a range of body colors (i.e., cream white, yellow, yellow-orange, golden, pink, red, etc.) and, while some of these color-mutations have been reported to be dominant, others have had incomplete dominant or even recessive effects. This could suggest that the

original color mutations have been influenced by additional genes in subsequent generations of different red tilapia varieties, possibly as a result of including genes from other tilapia species.

The breeding population of EGYPT red tilapia has considerable additive genetic variation of the degree of black spots both due to a large phenotypic variation and a moderate h^2 (0.24 ± 0.04). The h^2 of black spots was even higher in the last generation (0.62 ± 0.07), when the number of categories was increased. [71] reported that the black spots of individual Stirling red tilapia could be extremely reduced at certain stages (ages), but then return to its original state or even become more saturated afterwards. This could reflect the importance of melanin for photoprotection, camouflage and visual communication of tilapia during courting/brooding [67]. The scoring of black spots in the present breeding program was also significantly influenced by the time of recording, and the genetic correlation between black spots at mid-recording and harvest (0.8) stresses the necessity of recording black spots at the correct age and body weight to optimize genetic gain.

While red spots has been reported to be a problem in Stirling red [71], any yellow-orange-red pigmentation is welcomed in EGYPT red tilapia. Observations in the present breeding population suggested that most red pigmentation (erythrophores) were located in a layer between that of black spots (melanophores) and yellow pigmentation (xanthophores), and a majority of EGYPT red tilapia were classified as having a "red dominating over yellow" skin/scale color. The additive genetic variation of the skin/scale color was limited, however, both due to a small phenotypic variation and a surprisingly low h^2 (0.14 ± 0.02). The low h^2 could reflect difficulties scoring the trait, since the external color of fish is strongly influenced by the ratio of penetrating and substrate reflected light (albedo) in the grow-out environment [67]. Periods of strong sunshine, clear water and a dark substrate (large albedo) induce morphological darkening, while turbid water and cloud cover (small albedo) induce lighter external colors. Alternatively, the h^2 could reflect an additional additive genetic variation than that caused by the original colormutations, since no naturally colored tilapia are represented in the breeding population. The additive genetic variation of pigmented area was considerably larger, due to a large phenotypic variation and a high h^2 (0.51 ± 0.03). Although, the mean scores of skin/scale color and pigmented area were both significantly influenced by the time of recording, very high genetic.

I hybridized some EGYPT red fish with Stirling red fish. I noticed that the F1 offspring show vivid red blotches against an overall red background. It may be that in this hybrid the usual black blotches have turned crimson – potentially a commercially valuable color morph. The dashed line (black) includes all data collected in the growth trial experiment. The dotted line and points (red) include the F line data collected for selection purposes. The solid line and points (blue) include the corresponding M line data. It was noted in a previous report that when the original weight-thickness data were collected, the EGYPT fish were thicker and had higher fillet yields than the contract fish measured at that time.

In (Figure 21), the dashed line (black) includes all data in the growth trial experiment. The dotted line and points (red) include the F line data collected late in 2003 for selection purposes. The solid line and points (blue) include the corresponding M line data. It was noted in a previous report that when the original weight-thickness data were collected the EGYPT fish were thicker and had higher fillet yields than the contract fish measured at that time.

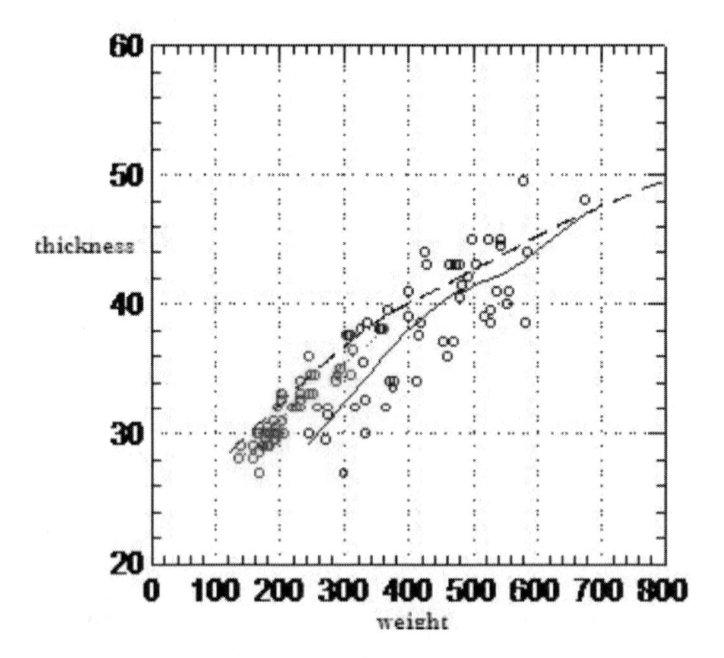

Figure 21. Body thickness plotted against weight in old and recent data.

Hatchery-Based Selection

The original selection plan was based on mass selection of two lines, "M" and "F" which are hybridized each generation to prevent inbreeding during production. This plan requires that individuals be fin-clipped to identify lines. Fin clipping is a limiting factor which prevents selection intensities from being stronger than 20% or so. The new hatchery system permits a more sophisticated and powerful selection program, as shown in Figure (22) below. Batches produced every month or two will generate overlapping sets of breeders with an effective population size so large that the inbreeding will be negligible. Selection intensities will be less than 10% and very possible less than 1%. The breeders already selected in the original programme will be rolled over into the new program without sacrificing any of the genetic gains already achieved.

Hatchery Performance As a Limiting Factor

The "dynamic" flow diagram for calculating breeder requirements and predicting selection is presented in Figure 22. The consensus figures which reflect the current performance of the new hatchery are shown below in an illustration of the Excel spreadsheet (Figure 22). It turns out that the system as a whole is highly sensitive to the biological parameters of hatchery performance. Changes of only ± 10% in spawning success and the eggs/female ratio have a very large effect on the number of breeders required, breeder replacement rate and the selection intensities which can be achieved. Thus, the performance of the hatchery deserves a lot of management attention.

Standardized Hatchery/Selection Procedures and Reporting

In this breeding program I discussed the desirability of bringing the hatchery and selection activities under the ISO umbrella at EGYPT. This should help ensure that the genetic program maintains its momentum and that its value to the company is properly documented. To enable ISO reporting, genetics-related activities and the *anticipated outcome* of those activities must both be specified, so that exceptions can be recognized and dealt with in a timely way. I began constructing an Excel "selection template" which will serve four purposes: (1) specify the actual selection procedures in enough detail so that requirements for facilities and personnel can be recognized before it is too late, (2) predict on a monthly and yearly basis how many animals of various sizes will be available for selection and breeding, (3) document costs and genetic progress and (4) provide a framework for ISO monitoring. The template illustrated in Figure 23 is part of a long Excel spreadsheet in which the value of each row is calculated from the value of the row above it, using appropriate growth rate formulae and survival data. A new spreadsheet is initialized when each batch is first put in the troughs. Future weights and numbers are predicted on a daily and/or stage-specific basis all the way up to selection and harvest six to eight months later. As the development of the batch proceeds, projected data are updated with real data collected on a regular schedule. In this way the projection becomes increasingly accurate as the time of selection and breeding approaches. The regular comparison of projected and actual measurements will serve to alert managers of problems, and also constitutes part of the ISO monitoring.

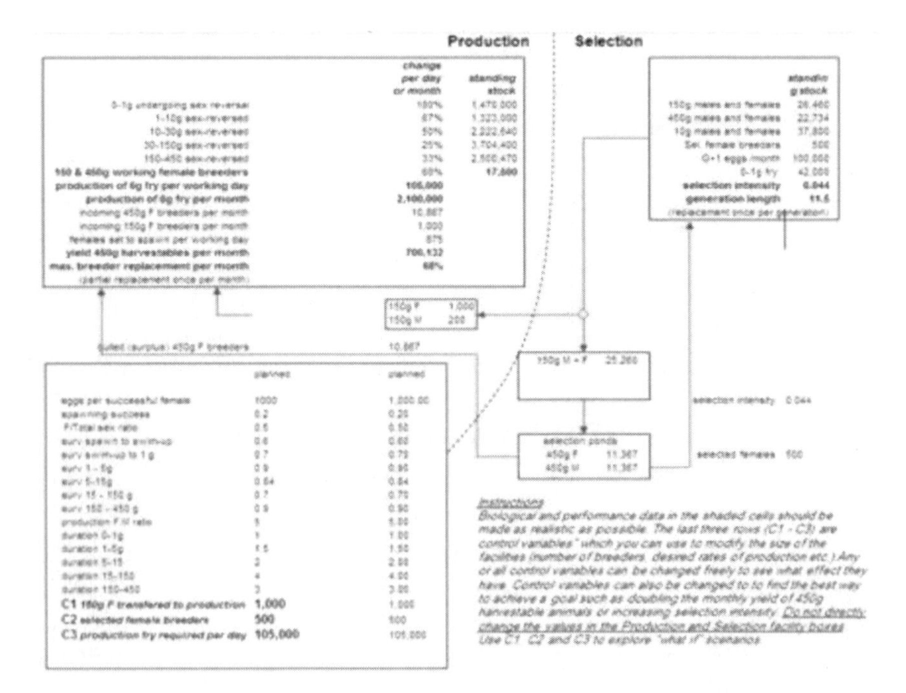

Figure 22. Updated hatchery and breeder spreadsheet.

Growth rate calculations will, initially, be strictly empirical using data collected during on the first batch to be run through the system, updated with data from subsequent batches. The daily growth rate (DGR) will be projected as well as estimated by multiple linear regressions, as follows:

(DGR) in each successive environment =
a_1 (regression constant)
+ a_2*T (temperature)
+ a_3*T^2
+ a_4*DGR (daily feed ration, possibly expressed as a percent of weight)
+ a_5*Prot (% protein in the ration)
+ a_6*T*DGR (interaction term, allowing for possible temperature effects on feeding)
+ a_7*W (weight, possibly with a transformation)
+ a_8*W^2
+ a_9*W*T (interaction term)
+ a_{10}*W*T*DGR (interaction term)
+ a_{11}*N (population density)

When initial estimates of $a_1 - a_{11}$ are available we will use the regression equation in an Excel spreadsheet to project the mean size of the animals from one day to the next. Thus we can get going with our predictions almost right away without doing any preliminary experiments. DGR is an empirical daily growth measure, not the parameter of a model such as the von Bertalanffy, but the operational definition of DGR is still open. When we have the data we can try several definitions and see which works best. The simplest definition of DGR is just (final weight – initial weight)/days. A somewhat more satisfactory definition may be instantaneous growth rate (ln final weight – ln initial weight)/days.

Status of the Growth Trials

The EGYPT growth trials use the reference strain or internal control technique, in which the growth of each test strains is compared with the growth of a reference strain in the same pond or tank. The EGYPT silver strain was used as a reference for this set of experiments.

stage, event or process	male	female	container	projected	actual	working value
spawn date						
female count			raceway			
spawning success						
fecundity						
total egg count	jar 1	jar 2	2 @ jar			
start transfer trough date						
end transfer trough date						
swimup count						
1st date	trough 1	trough 2	2 @ trough			
1st nursery(1) count			2 @ trough			
1st nursery(1) weight			2 @ trough	0.02		
1st mean temp			2 @ trough			
1st feed rate			2 @ trough			
2nd date			2 @ trough			
2nd nursery(1) count			2 @ trough			
2nd nursery(1) weight			2 @ trough			
2nd mean temp			2 @ trough			
.						
.						
.						
transfer concrete tank date						
1st date	tank 1		1 @ square concrete			
1st nursery(3) count			1 @ square concrete			
1st nursery(3) weight			1 @ square concrete			
1st mean temp			1 @ square concrete			
1st feed rate			1 @ square concrete			
2nd date			1 @ square concrete			
2nd nursery(3) count			1 @ square concrete			
2nd nursery(3) weight			1 @ square concrete			
.						
.						
.						
transfer pond date	pond 1		1@ 1-acre ponds			
1st date			1@ 1-acre ponds			
1st phase 1(pre-sex) count			1@ 1-acre ponds			
1st phase 1(pre-sex) weight			1@ 1-acre ponds			
1st mean temp			1@ 1-acre ponds			
1st feed rate			1@ 1-acre ponds			
2nd date			1@ 1-acre ponds			
2nd phase 1(pre-sex) count			1@ 1-acre ponds			
.						
.						
.						
6th feed rate			1@ 1-acre ponds			
7th date			1@ 1-acre ponds			
7th phase 2(sexed) count			1@ 1-acre ponds			
7th phase 2(sexed) weight			1@ 1-acre ponds			
7th mean temp			1@ 1-acre ponds			
7th feed rate	450	450	1@ 1-acre ponds			

Figure 23. Selection template for projection and evaluation.

Growth Rate of the EGYPT Silver Reference Strain

The absolute growth rate of the silver reference strain is shown in Figure 24. It can be seen that the fish grew a lot faster in the tank than in the pond environment. The point on the graph labeled "Tilapia" represents the growth rate claimed by a commercial seed supplier in Egypt in a tank growout system. The upper, dashed line is the projected growth rate of the Stirling fish in the EGYPT tanks, calculated by multiplying the growth of the EGYPT Stirling fish by 64% (see below). According to this calculation the performance of Stirling fish in the EGYPT tank environment should be as good as a (presumably) top-quality system elsewhere. Each point on the graph represents the mean weight of 30 – 50 fish.

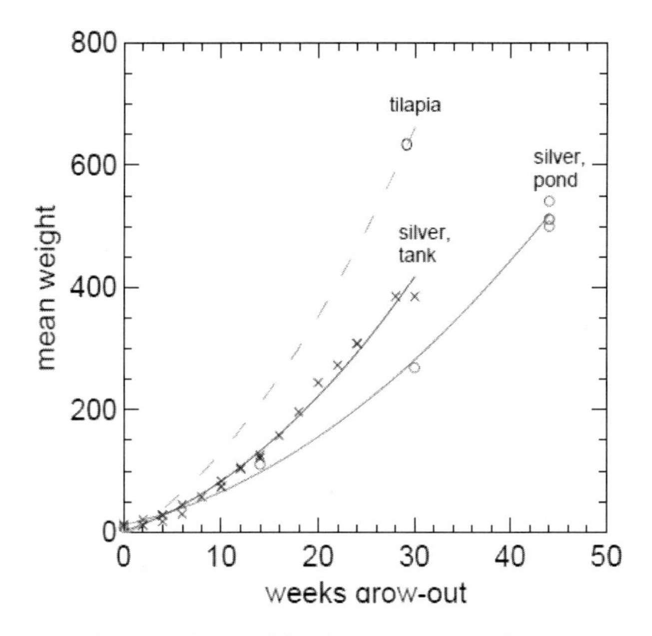

Figure 24. The absolute growth rate of the silver reference strain.

Relative Growth of the Taiwan Strain

The growth of the Taiwan is 22% faster than the growth of the silver strain averaged over the whole experiment. However, it should be noted that the two growth rates only begin to diverge when the fish weight approximately 100 grams, which is about the time they are transferred from the nursery to the G2 grow-out ponds. Each point on the graph represents the mean weight of 30 – 50 fish.

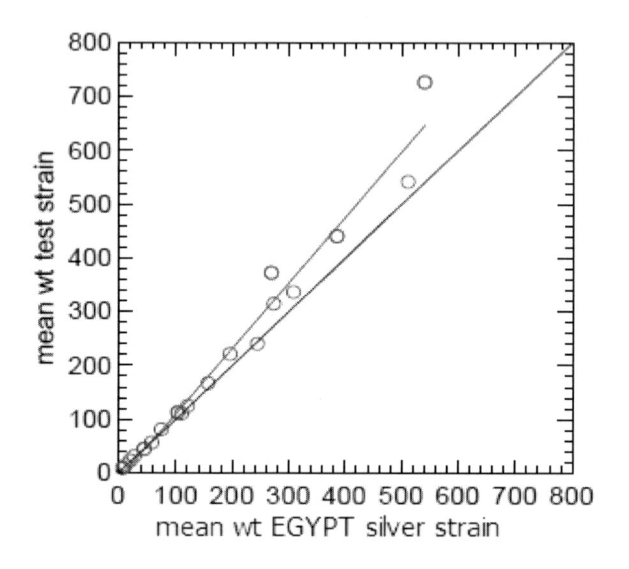

Figure 25. Relative growth of the Taiwan strain.

Relative Growth of the EGYPT Red Strain

The EGYPT red strain grows 17% faster than the Silver strain. The growth rates of the Taiwan and EGYPT red (22% and 17% respectively) are not significantly different. As more data come in this 5% difference may prove to be significant.

Relative Growth of the Stirling Red Strain

The Stirling strain grows 48% faster than the EGYPT silvers, on average. However, when the single low point is deleted from the analysis (the point obtained in the nursery pond), the relative growth of the Stirling fish is 64% faster than the reference strain. Except for the single nursery point these data were all obtained in the tank grow-out environment.

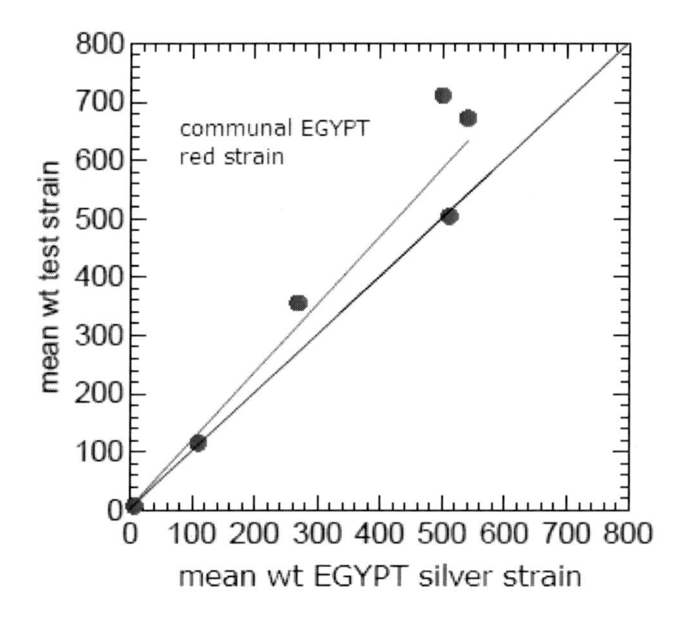

Figure 26. Relative growth of the EGYPT red strain.

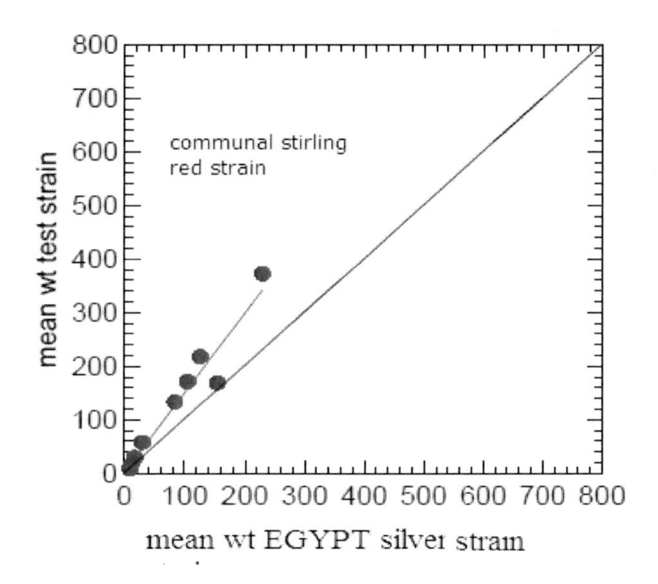

Figure 27. Relative growth of the Stirling red strain.

Stirling vs. Taiwan Growth in Tanks

The relative growth of the Stirling strain is higher than the Taiwan strain in the resource-rich tank environment (64% and 14% higher than the silver strain, respectively). I do not yet know whether it is also higher in the resource-rich G2 pond environment. Nor do I know whether it is higher than EGYPT red in tanks.

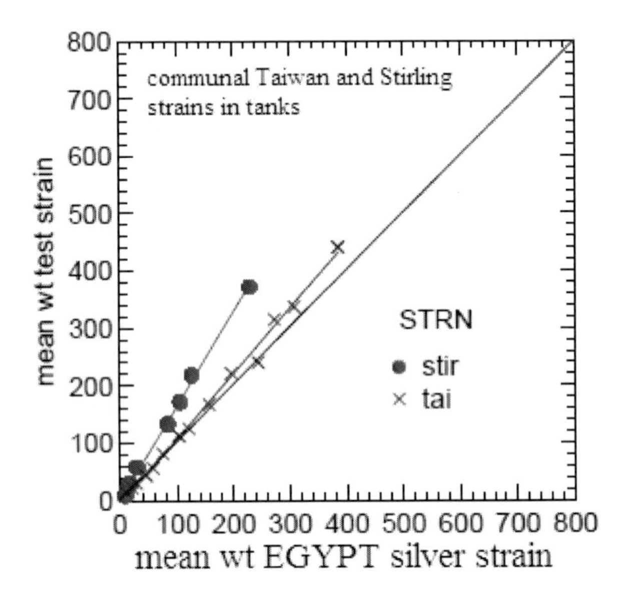

Figure 28. Stirling vs. Taiwan growth in tanks.

Predicted Fillet Yield of the Stirling Strain

Thickness measured at the "shoulder" of the fish is a good indicator of fillet yield for a fish of a given size. Data *from pond grow-out* were used to generate an equation which is shown as a dashed line in Figure 29. Measurements show that the EGYPT silver reference fish (open circles) grown in tanks are thicker than EGYPT fish grown in ponds. The Stirling fish are at least as thick as the EGYPT reference strain in the same environment, which suggests that at present we need not be concerned about the potential fillet yield of the Stirling strain. (Note the log scale on the X axis.)

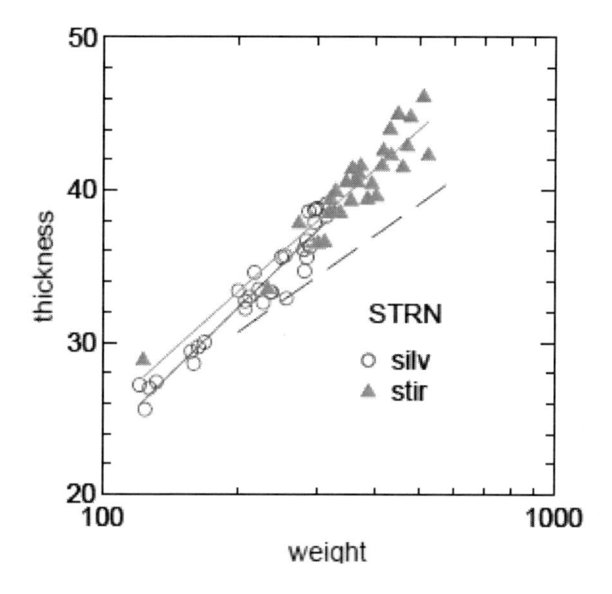

Figure 29. Predicted fillet yield of the Stirling strain.

Preliminary Conclusions from the Growth Trials

It appears that the EGYPT reds and Taiwan reds grow approximately 17% and 22% faster, respectively, than the EGYPT silver reference fish in G2 ponds. Taiwan may grow 5% faster than EGYPT reds in this environment but the difference is not yet statistically significant. More experiments would be needed before one would decide to replace the current EGYPT red strain with Taiwan fish for pond growth. The relative growth of the Stirling strain in G2 grow-out ponds is not yet known but the experiment is scheduled. The observed superiority of the Stirling over the Taiwan fish in tanks is very great, 64-14=50%. A direct test of the growth of this strain relative to EGYPT reds in tanks is scheduled. None of the strains grow significantly faster than the silver reference strain in nursery ponds, which suggests that this environment may severely limit growth (i.e., growth is set by the rate of input of food, not by the genotype's ability to grow in the presence of unlimited food). The results of the experiments so far are encouraging and suggest that the Stirling strain is superior to the other strains in the tank environment. Whether this observation can be replicated and whether it extends to the G2 grow-out environment remains to be seen.

Age and Size at Selection

The question whether or not to select large animals early, e.g. when they are moved out of the nursery ponds, was raised again during this visit. The two sides of the argument for early selection are: (*Pro)* early selection will shorten the generation interval and speed up the rate of genetic gain by 50% or more; (*Con*) early selection may distort the growth curve, giving rise to rapid maturation and a fish that grows slowly after maturation. There is considerable anecdotal and some experimental evidence that this happens in tilapia, where the phenomenon is called "hitting the wall". Experiments to settle the issue at EGYPT by looking at the correlation between early and late grows rates have been planned but not yet carried out. The interim decision made during this visit was *not* to select our future G_1 broodstock as they are leaving the nursery, but to transfer them to the G2 ponds and wait until they approximately harvest size before selecting them. This decision was made partly because the whole strain may be replaced anyway and partly because it was felt that the selection intensity/generation time issue should be solved in some other way, i.e., by changing the marking system.

Selection Intensity Will Limit the Rate of Genetic Gain

In the existing program the main factor which limits the rate of genetic progress is the number of fish which can be marked by fin-clipping, which is between four and six thousand. The estimated number of selected breeders required to replace breeders annually is about 300 for each of the two lines. Mortality between tagging and selection will reduce the number available for selection, with the result that the selection intensity is anticipated to be approximately 10%. This is actually rather good in relation to what can be achieved in terrestrial animals which have only a few offspring. However, it is far less than what could be achieved in tilapia if a different marking technique could be used. The relationship between selection intensity and the expected rate of genetic gain is non-linear. It can be calculated that, other genetic factors being equal, increasing the rate of genetic gain by 50% would require us to decrease the fraction of animals selected in each generation from 10% down to approximately 0.42% – which would mean marking 95,000 animals instead of 4,000. This is clearly impossible with fin clips. However, it would be easy if the fish had natural colour marks such as the dominant red gene of the Stirling red strain. (Increasing the selection intensity enough to *double* the currently projected rate of genetic gain would mean marking about 700,000 fish instead of 4,000.)

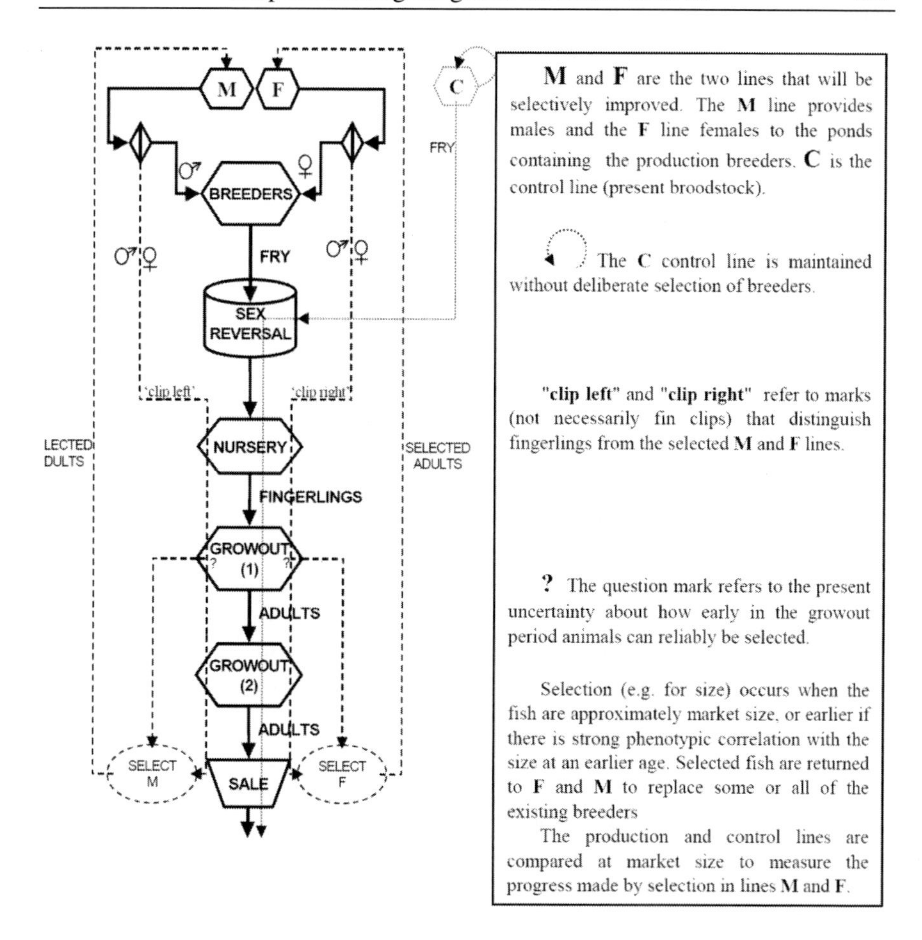

Figure 30. Selection flow chart.

Rate of Broodstock Replacement Will Limit the Rate of Genetic Gain

The present intention is to maintain a stock of 300 selected female breeders in each of the M and F lines. These are called "grandparent" breeders and are labeled *M* and *F* in the flow diagram (Figure 30.). With 300 breeders selected out of 4,000 – 6,000 marked it may be possible to maintain a selection intensity of 10% while replacing all the breeders every year to minimize the generation time. However, 300 breeders will not produce enough fry to replace the breeders of the parental generation, which are used to generate the F1 hybrid offspring for the production line. It appears that 30,000 breeders will be needed annually but 300 breeders will produce only about 12,000 fry/month, which is not enough when mortality is taken into consideration. It was suggested that the parental breeders could be replaced every second year instead of every year, but this represents a delay in the transfer of the benefits of genetic improvement to the production line. A better solution is to find a marking system which allows a larger number of fish to be selected than the present fin-clipping system.

Fish Marking Procedure and Selection Intensities

Fin clipping to distinguish between strains

The fin-clipping procedure has been perfected at EGYPT. The rate of loss of marks by regeneration is quite low and the marked fish are easy to spot on the sorting tables. The maximum number of fish that can be marked in one episode, i.e., one round of selection, is estimated to be 4,000 to 6,000. This establishes an upper limit to the selection intensity which can be achieved with this marking procedure to approximately 10%.

Natural colour marks (red vs. silver wild-type)

Much higher selection intensities could be achieved if the presence or absence of the dominant red gene can be used instead of left/right fin clipping to distinguish M and F strains. There is essentially no upper limit to the selection intensity except the number of fish which can be measured, which would be very high if they are pre-screened to concentrate the larger ones before measurement. The F1 hybrids between the homozygous red and homozygous silver strains would themselves be red, as required for production. These F1 red fish mated among themselves would produce 25% silver fish in each generation, i.e., would not breed true.

P.I.T. tag markers and DNA samples

When each new generation of selected "grandparent" broodstock is produced the animals which are chosen to become breeders should be P.I.T. tagged so they can be identified and removed from the breeder pool at the

> appropriate time. Tissue samples should be taken from the fins at the same time as the fish are tagged and carefully stored for later DNA analysis. The DNA data will be at some point used for several purposes: monitoring the rate of inbreeding and loss of genetic diversity (especially important if the Stirling strain is used), documentation of the development of the broodstock for forensic and legal protection of EGYPT's proprietary strain. A good source of PIT tags, readers and injectors is Biomark Inc.

In a study by [40] External color traits in the present breeding program were all defined and analyzed as polygenic traits. Most populations of red tilapia have individuals with different degrees of black spots, which negatively affect their appearances. [72] reported that up to 25% of the body surface of Stirling red tilapia could be covered with black spots. Studies have suggested that several color-mutations are dominant over the natural color of Nile tilapia [73] and Mozambique tilapia [74; 75], but the offspring usually had variable degrees of black spots. These observations could suggest an incomplete dominant inheritance of the color mutations or that black spots also are governed by additional genes and environmental factors. Since, inbreeding increases the expression of recessive genes, a recessive inheritance could explain the continued problem with black spots in many hatchery populations of red tilapia. Here are the size distributions for the colour morphs. To the eye, they seem different but the Analysis of Variance Table shows that they are not (p=.09) nor are any of the pairwise differences significant (p=1.0, 0.229, 0.345 etc.) This is also true after log transformation to reduce the right-hand skewness of the data. The results of color morhps occurrence and size distribution of different color morphs are presented in Figures (32, 33, 34, 35, 36, 37, 38 and 39) All available growth data for EGYPT reds plotted against EGYPT silver reference (Figure 40) fish growing in the same tank or pond (mean weights, not individual weights). This is a pool of data collected over several years. Note that the G2 reds (blue points), at less than 50 grams, are still too small for a meaningful comparison with the ordinary reds (red points). The dotted black line indicates equal growth rates for the reds and silvers. On the whole, the EGYPT reds have not done as well as the silvers (they fall below the equal-growth line). In the graph below the EGYPT reds and Stirling reds are plotted against matched batches of EGYPT silvers. In other words, when points are above the diagonal line it indicates that the fish in question are growing faster than silvers. My interpretation of the current results this year is that grow-out conditions were poor (fish are thin) and the genetic growth superiority of EGYPT reds over EGYPT silvers was not expressed.

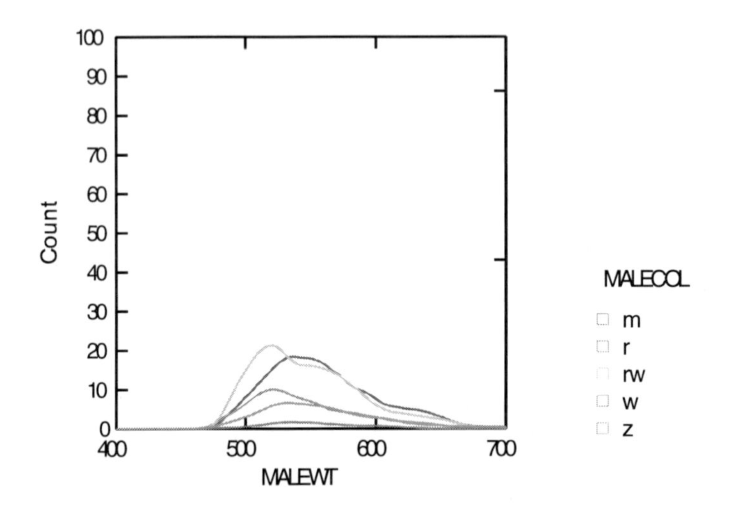

Figure 31. The size distributions for the colour morphs.

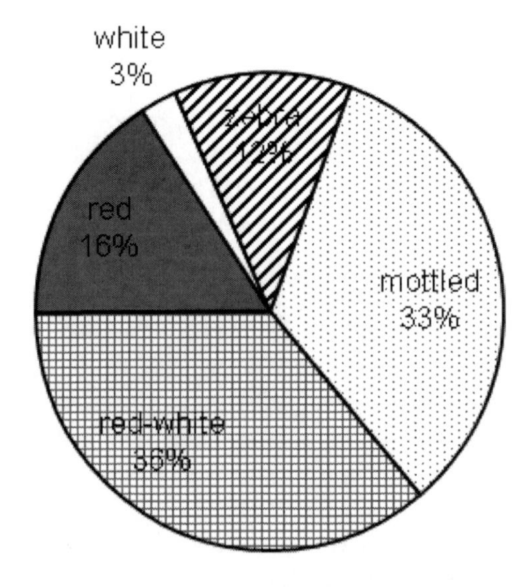

Figure 32. Occurance of color morphs in selected male G2 fish.

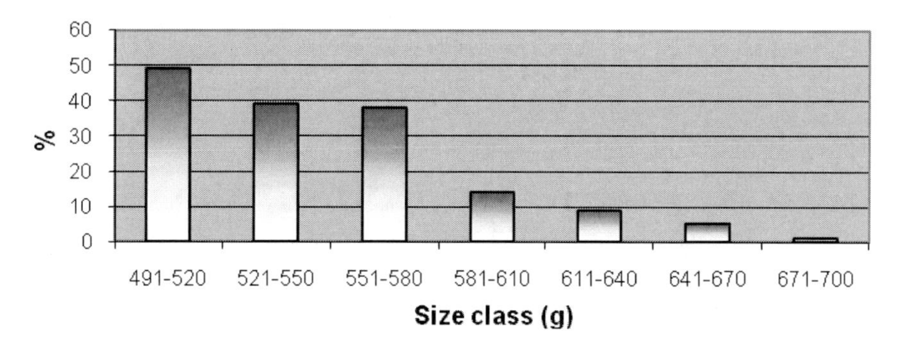

Figure 33. Size distribution of EGYPT red G2 “Red-white” males.

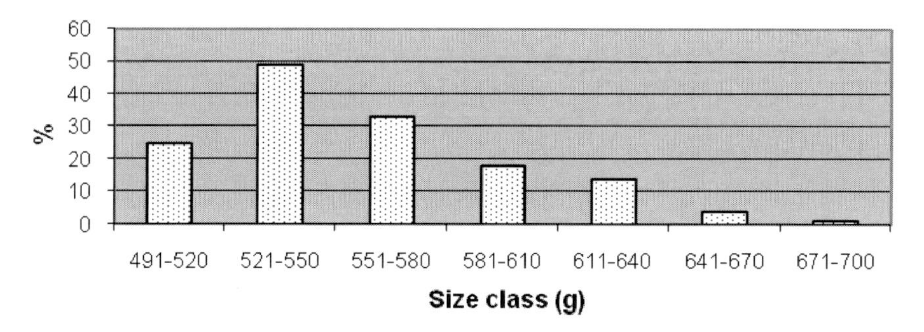

Figure 34. Size distribution of EGYPT red G2 “Mottled” males.

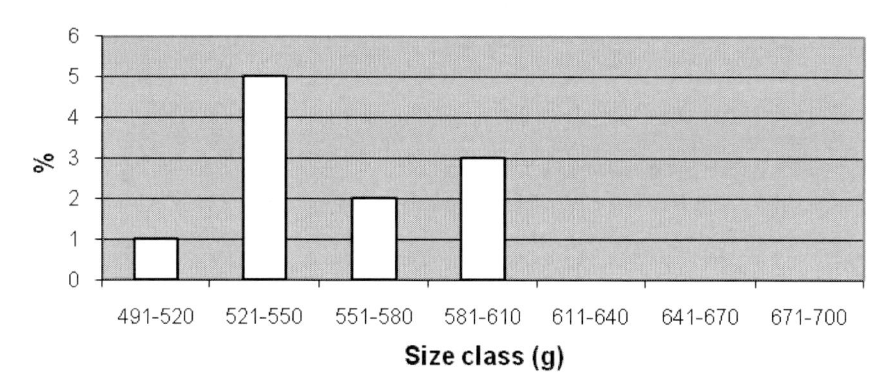

Figure 35. Size distribution of EGYPT red “White” males.

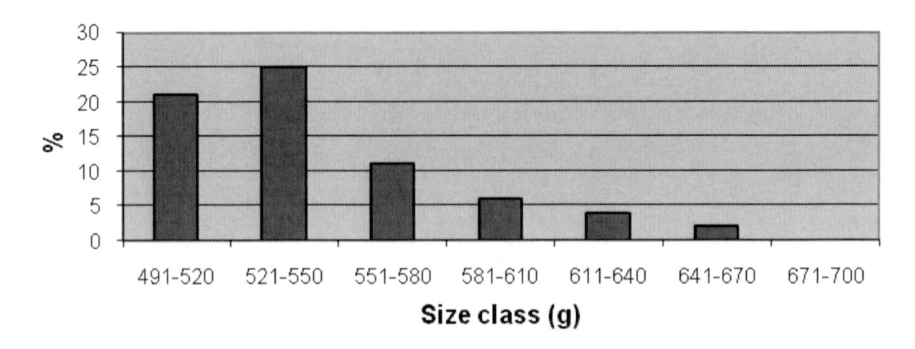

Figure 36. Size distribution of EGYPT red G2 Zebra males.

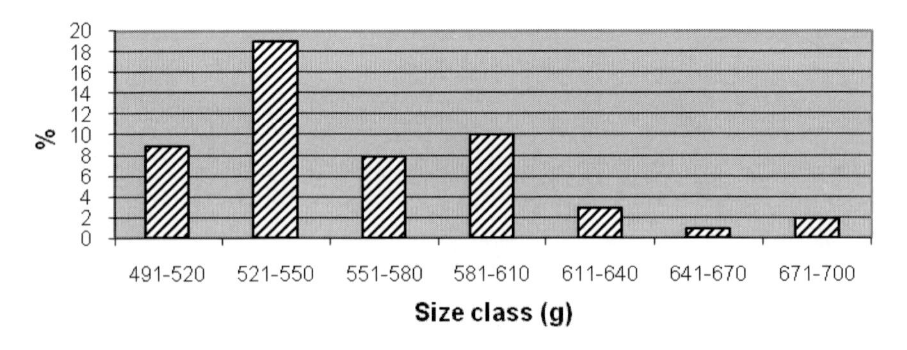

Figure 37. Size distribution of EGYPT red G2 "Zebra" males.

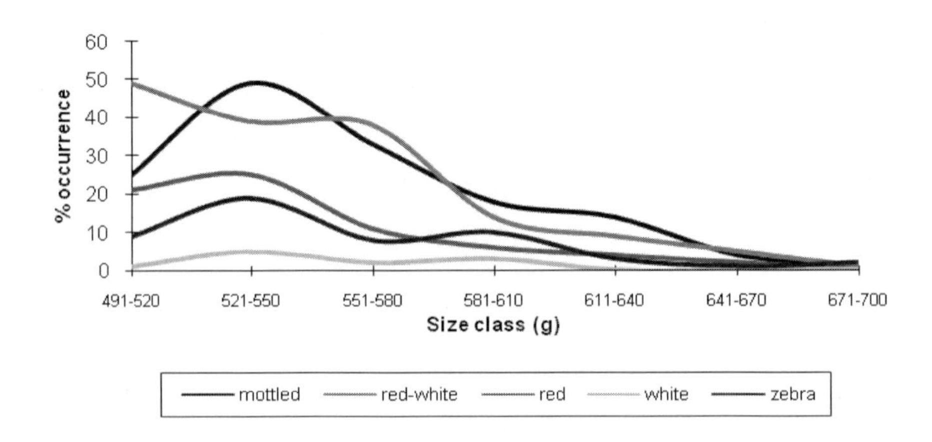

Figure 38. Size distribution of EGYPT red G2 "males color morphs:.

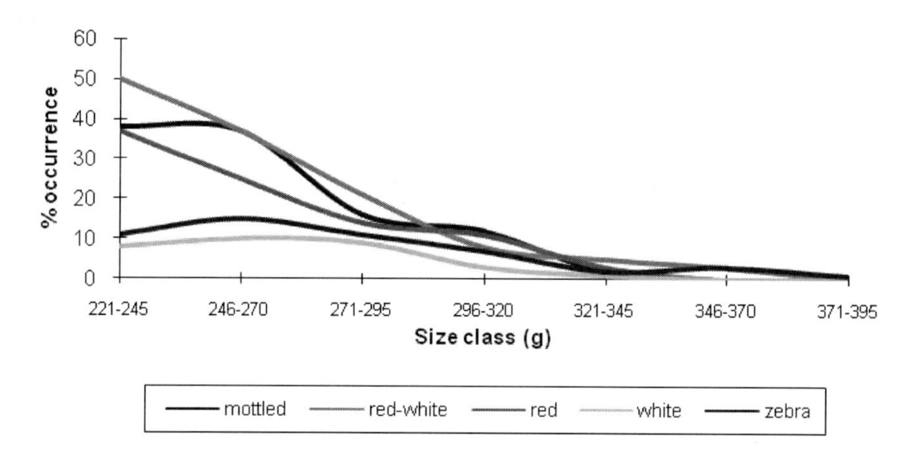

Figure 39. Size distribution of EGYPT red G2 female color morphs.

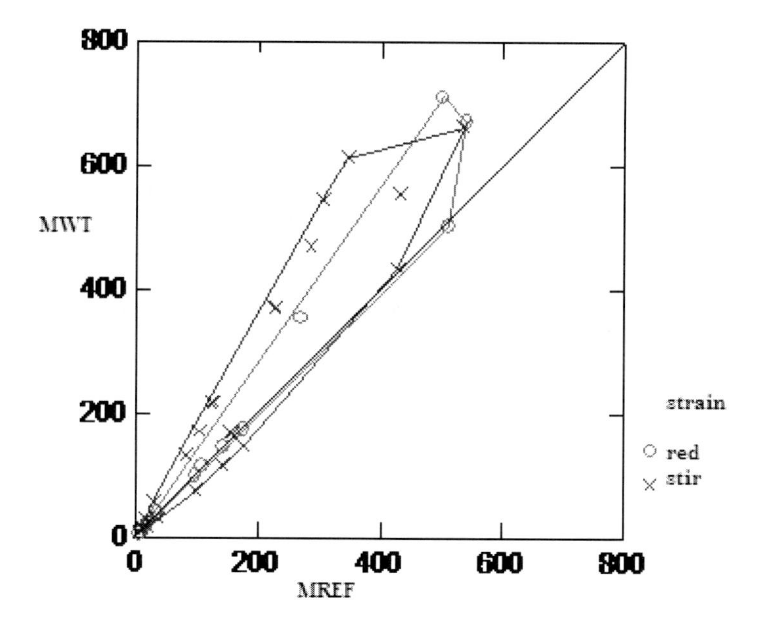

Figure 40. growth data for EGYPT reds plotted against EGYPT silver reference.

Conclusion and Recommendation for the Breeding Program

The breeding objective was to increase fillet yield as a correlated response to such selection. The results of this breeding program for fillet weight indicate high correlation between body weight and fillet weight, which means that the genetic improvement of one of the traits without achieving a proportional genetic change in the other is difficult or even impossible. So, fillet yield is expected to increase with increasing body weight, both genetically and phenotypically. Estimated heritabilities for fillet yield in Nile tilapia population ranged from moderate to high suggesting rapid direct responses to selection. Family common environmental effects were more important for body weight and for fillet weight. Considering fillet weight as target traits, body weight can be used as the selection criterion. It is concluded that it is possible to genetically improve fillet weight of Nile tilapia.

This colour business is a pain. Generally speaking pure reds which have the colour-determining system characteristic do not grow as well as mottled fish. Furthermore they have lower survival (which is why they're rare) and fecundity. We have always known that. The growth trials seem to have shown that mottled EGYPT reds grow a bit better than EGYPT silvers when conditions are good. When conditions are very good we had a signal – never verified – that the Stirling pure reds did exceptionally well, but they are pure niloticus and have a different colour-determining system.

"My conclusion [from the growth trials] is that:

1) Under food limiting or otherwise poor conditions the Stirling, EGYPT red and silver grow at about the same rate.

2) Under the usual conditions the Stirling and EGYPT red grow 15% – 20% faster than the silvers.

3) Under exceptionally favourable conditions the Stirling are capable of performing 30% - 40% better than the EGYPT reds."

References

[1] Eknath, A.E., Acosta, B.O, Genetic Improvement of Farmed Tilapias (GIFT) Project: Final Report, March 1988 to December 1997.

International Center for Living Aquatic Resources Management, Makati City, Philippines. (1998).

[2] Rutten, M.J.M., Komen, H., Bovenhuis, H., Longitudinal genetic analysis of Nile tilapia (Oreochromis niloticus L.) body weight using a random regression model. *Aquaculture* 246, 101–113. (2005a).

[3] Rutten, M.J.M., Bovenhuis, H., Komen, H., Genetic parameters for fillet traits and body measurements in Nile tilapia (Oreochromis niloticus L.). *Aquaculture* 246, 125–132. (2005b).

[4] Ponzoni, R.W., Hamzah, A.B., Saadiah, T., Norhidayat, K., Genetic parameters and response to selection for live weight in the GIFT strain of Nile tilapia (Oreochromis niloticus). *Aquaculture* 247, 203–210. (2005).

[5] Maluwa, A.O., Gjerde, B., Ponzoni, R.W., Genetic parameters and genotype by environment interaction for body weight of Oreochromis shiranus. *Aquaculture* 259, 47–55. (2006).

[6] FAO, FAO yearbook. *Fishery and Aquaculture Statistics,* 2008. 74 pp. (2010).

[7] Fitzsimmons, K., Martinez-Garcia, R., Gonzalez-Alanis, P., Why tilapia is becoming the most important food fish on the planet. In: Liu, L.P., Fitzsimmons, K. (Eds.), *Proceedings of the 9th International Symposium on Tilapia in Aquaculture.* Shanghai, pp. 1–8 (409 pp.). (2011).

[8] Chervinski, J., Environmental physiology of tilapias. In: Pullin, R.S.V., Lowe-McConnell, R.H. (Eds.), The Biology and Culture of Tilapias: ICLARM Conference Proceedings, 7, pp. 119–128. 432 p. (1982).

[9] Lovshin, L.L., Tilapia hybridization. In: Pullin, R.S.V., Lowe-McConnell, R.H. (Eds.), *The Biology and Culture of Tilapias: ICLARM Conference Proceedings,* 7, pp. 279–308. 432p. (1982).

[10] Maclean, J.L., 1984. Tilapia - the aquatic chicken. ICLARM Newsletter, 7 (1): 17. Palada-de Vera, M.S. and Eknath, A.E., 1993. Predictability of individual growth rates in tilapia. *Aquaculture,* Ill: 147-158.

[11] Smith, I.R. and Pullin, R.S.V., Tilapia production booms in the Philippines. *ICLARM Newsletter,* 7 (1): 7-9. (1984).

[12] Pullin, R.S.V., Tilapias: 'everyman's fish'. *Biologist,* 32(2): 84-88. (1985).

[13] Pullin, R.S.V. (Editor), 1988. Tilapia genetic resources for aquaculture. ICLARM Conference *Proceedings 16. International Center for Living Aquatic Resources Management, Manila*, Philippines, 108 pp. (1985).

[14] Pullin, R.S.V., Capili, J.B., Genetic improvement of tilapias: problems and prospects. In: Pullin, R.S.V., Bhukaswan, T., Tonguthai, K.,

Maclean, J.L. (Eds.), The Second International Symposium on Tilapia in Aquaculture. *ICLARM Conf. Proc,* 15, pp. 259–266. (1988).

[15] El-Sayed, A.F.M., 2006. Tilapia Culture. *CABI Publishing,* Wallingford (i-xvi+277 pp.).

[16] Fessehaye, Y., El-bialy, Z., Rezk, M.A., Crooijmans, R., Bovenhuis, H., Komen, H., 2006. Mating systems and male reproductive success in Nile tilapia (Oreochromis niloticus) in breeding hapas: a microsatellite analysis. *Aquaculture* 256, 148–158.

[17] Li, S.F., He, X.J., Hu, G.C., Cai,W.Q., Deng, X.W., Zhou, P.Y., 2006. Improving growth performance and caudal fin stripe pattern in selected F6–F8 generations of GIFT Nile tilapia (Oreochromis niloticus L.) using mass selection. *Aquac. Res.* 2006, 1–7.

[18] Zhao, J.L., 2011. Tilapia germplasmin China: chance and challenge. In: Liu, L.P., Fitzsimmons, K. (Eds.), *Proceedings of the 9th International Symposium on Tilapia in Aquaculture.* Shanghai, pp. 217–221. 409 pp.

[19] FAO, 2005–2009. Cultured Aquatic Species Information Programme Oreochromis niloticus (Linnaeus, 1758). In: Rakocy, J.E. (Ed.), FAO Fisheries and Aquaculture Department [online]. Rome. Available at: http://www.fao.org/fishery/culturedspecies/Oreochromis_niloticus/en# tcN900FE (accessed: 2.04.2010).

[20] Bentsen, H.B., Gjerde, B., Nguyen, N.H., Rye,M., Ponzoni, R.W., Palada de Vera,M.S., Bolivar, H.L., Velasco, R.R., Danting, J.C., Dionisio, E.E., Longalong, F.M., Reyes, R.A., Abella, T.A., Tayamen, M.M., Eknath, A.E., 2012. Genetic improvement of farmed tilapias: genetic parameters for body weight at harvest in Nile tilapia (Oreochromis niloticus) during five generations of testing in multiple environments. *Aquaculture* 338–341, 56–65.

[21] Neira, N., Breeding in aquaculture species: genetic improvement programs in developing countries. *9th World Congress on Genetics Applied to Livestock Production,* Leipzig, Germany, p. 8. (2010).

[22] Ponzoni, R.W., Nguyen, N.H., Khaw, H.L., Hamzah, A., Bakar, K.R.A., Yee, H.Y., Genetic improvement of Nile tilapia (Oreochromis niloticus) with special reference to the work conducted by the World Fish Center with the GIFT strain. *Reviews in Aquaculture* 3, 27–41. (2011).

[23] Henryon, M., Jokumsen, A., Berg, P., Lund, I., Pedersen, P.B., Olesen, N.J., Slierendrecht, W.J., Genetic variation for growth rate, feed conversion efficiency, and disease resistance exists within a farmed population of rainbow trout. *Aquaculture* 209, 59–76. (2002).

[24] Dumas, A., France, J., Bureau, D., Modelling growth and body composition in fish nutrition: where have we been and where are we going? *Aquaculture Research* 41, 161–181. (2010).

[25] Jobling, M., The thermal growth coefficient (TGC) model of fish growth: a cautionary note. *Aquaculture Research* 34, 581–584. (2003).

[26] Rezk, M.A., Ponzoni, R.W., Khaw, H.L., Kamel, E., Dawood, T., John, J., Selective breeding for increased body weight in a synthetic breed of Egyptian Nile tilapia, Oreochromis niloticus: response to selection and genetic parameters. *Aquaculture* 293, 187–194. (2009).

[27] Ponzoni, R.W., Khaw, H.L., Nguyen, H.N., Hamzah, A., Inbreeding and effective population size in theMalaysian nucleus of the GIFT strain of Nile tilapia (Oreochromis niloticus). *Aquaculture* 302, 42–48. (2010).

[28] Eknath, A.E., Hulata, G., Use of exchange of genetic resources of Nile Tilapia (Oreochromis niloticus). *Reviews in Aquaculture* 1, 197–213. (2009).

[29] Nguyen, N.H., Ponzoni, R.W., Abu-Bakar, K.R., Hamzah, A., Khaw, H.L., Yee, H.Y., Correlated response in fillet weight and yield to selection for increased harvest weight in genetically improved farmed tilapia (GIFT strain), Oreochromis niloticus. *Aquaculture* 305, 1–5. (2010).

[30] Kause, A., Paananen, T., Ritola, O., Koskinen, H., 2007. Direct and indirect selection of visceral lipid weight, fillet weight, and fillet percentage in a rainbow trout breeding program. *Journal of Animal Science* 85, 3218–3227.

[31] Einen, O., Mørkøre, T., Rørå, A.M.B., Thomassen, M.K., Feed ration prior to slaughter—a potential tool for managing product quality of Atlantic salmon (Salmo salar). *Aquaculture* 178, 149–169. (1999).

[32] Powell, J., White, I., Guy, D., Brotherstone, S., Genetic parameters of production traits in Atlantic salmon (Salmo salar). *Aquaculture* 274 (2–4), 225–231. (2008).

[33] Sang, N.V., Thomassen, M., Klemetsdal, G., Gjøen, H.M., Prediction of fillet weight, fillet yield, and fillet fat for live river catfish (Pangasianodon hypophthalmus). *Aquaculture* 288, 166–171. (2009).

[34] Saillant, E., Dupont-Nivet, M., Sabourault, M., Haffray, P., Laureau, S., Vidal, M.-O., Van Sang, N., Thomassen,M., Klemetsdal, G., Gjøen, H.M., Prediction of fillet weight, fillet yield and fillet fat for live river catfish (Pangasianodon hypophthalmus). *Aquaculture* 288, 166–171. (2009).

[35] Rye, M., Gjerde, B., Gjedrem, T., Genetic development programs for aquaculture species in developed countries. *9th World Congress on Genetics Applied to Livestock production,* Leipzig, Germany, August 1–6, p. 8. (2010).

[36] Megahed, (2014).

[37] Meyer, K., 2004. Scope for a random regression model in genetic evaluation of beef cattle for growth. *Livestock Production Science* 86, 69–83.

[38] Velasco, R.R., Janagap, C.C., De Vera, M.P., Afan, L.B., Reyes, R.A., Eknath, A.E., 1995. Genetic improvement of farmed tilapias: estimation of heritability of body and carcass traits of Nile tilapia (Oreochromis niloticus). *Aquaculture* 137, 280–281.

[39] Gjerde, B., Mengistu, S.B., Ødegard, J., Johansen, H., Altamirano, D.S., 2012. Quantitative genetics of body weight, fillet weight and fillet yield in Nile tilapia (Oreochromis niloticus). *Aquaculture* 342–343, 117–124.

[40] Thodesen, J. (Ma, D.Y.)., Rye, M., Wang, Y.X., Yang, K.S., Bentsen, H.B., Gjedrem, T., 2011. Genetic improvement of tilapias in China: genetic parameters and selection responses in growth of Nile tilapia (Oreochromis niloticus) after six generations of multi-trait selection for growth and fillet yield. *Aquaculture* 322–323, 51–64.

[41] Falconer, D.S., Mackay, T.F.C., 1996. Introduction to Quantitative Genetics, Fourth edition. Longman Group Limited, Harlow, Essex, U.K.. 464 pp.

[42] Gjedrem, T., 1992. Breeding plans for rainbow trout. A*quaculture* 100, 73–83.

[43] Bosworth, B.G., Holland, M., Brazil, B.L., 2001. Evaluation of ultrasound imagery and body shape to predict carcass and fillet yield in farm-raised catfish. *Journal of Animal Science* 79, 1483–1490.

[44] Rutten, M.J.M., Bovenhuis, H., Komen, H., 2004. Modeling fillet traits based on body measurements in three Nile tilapia strains (Oreochromis niloticus L.). *Aquaculture* 231, 113–122.

[45] Teichert-Coddington, D.R., Smitherman, R.O., 1988. Lack of response by tilapia Nilotica to mass selection for rapid early growth. *Trans. Am. Fish. Soc.* 117, 297–300.

[46] Huang, C.M., Liao, I.C., 1990. Response to mass selection for growth rate in O. niloticus. *Aquaculture* 85, 199–205.

[47] Bosworth, B.G., Libey, G.S., Notter, D.R., 1998. Relationships among total weight, body shape, visceral components, and fillet traits in Palmetto Bass (Stripped Bass female Morone saxatilis x White Bass

male M. chrysops) and Paradise Bass (Stripped Bass female M. saxatilis x Yellow Bass male M. mississippiensis). *J. World Aquac. Soc.* 29, 40–50.

[48] Bolivar, R.B., Newkirk, G.F., 2002. Response to within family selection for body weight in Nile tilapia (Oreochromis niloticus) using a single-trait animal model. *Aquaculture* 204, 371–381.

[49] Luan, T.D., 2010. Genetic studies of Nile tilapia (Oreochromis niloticus) for farming in Northern Vietnam: growth, survival and cold tolerance in different farm environments. *Philosophiae Doctor (PhD) Thesis* 2010:04. Norwegian University of Life Sciences (141 pp.).

[50] Neira, R., Lhorente, J.P., Araneda, C., Diaz, N., Bustos, E., Alert, A., 2004. Studies on arcass quality traits in two populations of Coho salmon (Oncorhynchus kisutch): phenotypic and genetic parameters. *Aquaculture* 241, 117–131.

[51] Navarro, A., Zamorano, M.J., Hildebrandt, S., Ginés, R., Aguilera, C., Afonso, J.M., 2009. Estimates of heritabilities and genetic correlations for growth and carcass traits in gilthead seabream (Sparus auratus L.), under industrial conditions. *Aquaculture* 289, 225–230.

[52] Havenstein, G.B., Ferket, P.R., Qureshi, M.A., 2003. Carcass composition and yield of 1957 versus 2001 broilers when fed representative 1957 and 2001 broiler diets. *Poultry Science* 82, 1509–1518.

[53] Gjedrem, T. (Ed.), 2005. Selection and Breeding Programs in Aquaculture. *Springer,* Dordrecht, The Netherlands. 364 pp.

[54] Thodesen, J., Grisdale-Helland, B., Helland, S.J., Gjerde, B., 1999. Feed intake, growth and feed utilization of offspring from wild and selected Atlantic salmon (Salmo salar). *Aquaculture* 189, 237–246.

[55] Charo-Karisa, H., Komen, H., Rezk, M.A., Ponzoni, R.W., van Arendonk, J.A.M., Bovenhuis, H., 2006. Heritability estimates and response to selection for growth of Nile tilapia (Oreochromis niloticus) in low-input earthen ponds. *Aquaculture* 261, 479–486.

[56] Blonk, R.J.W., Komen, J., Tenghe, A., Kamstra, A., van Arendonk, J.A.M., 2010. Heritability of shape in common sole, Solea solea, estimated from image analysis data. *Aquaculture* 307, 6–11.

[57] Kause, A., Ritola, O., Paananen, T., Eskelinen, U., Mäntysaari, E., 2003a. Big and beautiful? Quantitative genetic parameters for appearance of large rainbow trout. *Journal of Fish Biology* 62, 610–622.

[58] Gjerde, B., Schaeffer, L.R., 1989. Body traits in rainbow trout. II. Estimates of heritabilities and of phenotypic and genetic correlations. *Aquaculture* 80, 25–44.
[59] Kim, J.H., Lee, J.H., Kim, H.C., Noh, J.K., Kang, J.H., Kim, K.K., 2011. Body shape and growth in reciprocal crosses of wild and farmed olive flounder, Paralichthys olivaceus. *Journal of the World Aquaculture Society* 42, 268–274.
[60] Khaw, H.L., Bovenhuis, H., Ponzoni, R.W., Rezk, M.A., Charo-Karisa, H., Komen, H., 2009.Genetic analysis of Nile tilapia (Oreochromis niloticus) selection line reared in two input environments. *Aquaculture* 294, 37–42.
[61] Khaw, H.L., Ponzoni, R.W., Hamzah, A., Abu-Bakar, K.R., Bijma, P., 2012. Genotype by production environment interaction in the GIFT strain of Nile tilapia (Oreochromis niloticus). *Aquaculture* 326–329, 53–60.
[62] Moav, R., Hulata, G., Wohlfarth, G., 1975. Genetic differences between the Chinese and European races of the common carp. *Heredity* 34, 323–340.
[63] Kause, A., Ritola, O., Paananen, T., Mäntysaari, E., Eskelinen, U., 2003b. Selection against early maturity in large rainbow trout Oncorhynchus mykiss: the quantitative genetics of sexual dimorphism and genotype-by-environment interactions. *Aquaculture* 228, 53–68.
[64] Sylvén, S., Rye, M., Simianer, H., 1991. Interaction of genotype with production system for slaughter weight in rainbow trout (Oncorhynchus mykiss). *Livestock Production Science* 28, 253–263.
[65] Hutchings, J.A., 2011. Old wine in new bottles: reaction norms in salmonid fishes. *Heredity* 106, 421–437.
[66] Dupont-Nivet, M., Chevassus, B., Mauger, S., Hafray, P., Vandeputte, M., 2010. Side effects of sexual maturation on heritability estimates in rainbow trout (Oncorhynchus mykiss). *Aquaculture Research* 41, e878–e880.
[67] Leclercq, E., Taylor, J.F., Migaud, H., 2010. Morphological skin colour changes in teleosts. *Fish Fish.* 11, 159–193.
[68] Wohlfarth, G.W., Rothbard, S., Hulata, G., Szweigman, D., 1990. Inheritance of red body coloration in Taiwanese tilapias and in Oreochromis mossambicus. *Aquaculture* 84, 219–234.
[69] Reich, L., Don, J., Avtalion, R.R., 1990. Inheritance of the red color in tilapias. *Genetica* 80, 195–200.

[70] Huang, C.M., Chang, S.L., Cheng, H.J., Liao, I.C., 1988. Single gene inheritance of red body coloration in Taiwanese red tilapia. *Aquaculture* 74, 227–232.

[71] Rajaee, A.H., 2011. Genetic approaches to the analysis of body colouration in Nile tilapia (Oreochromis niloticus L.). Doctor of Philosophy Thesis. University of Stirling (190 pp.).

[72] McAndrew, B.J., Roubal, F.R., Roberts, R.J., Bullock, A.M., McEwen, I.M., 1988. The genetics and histology of red, blond and associated colour variants in Oreochromis niloticus. *Genetica* 76, 127–137.

[73] Moreira, A.A., Moreira, H.L.M., Hilsdorf, A.W.S., 2005. Comparative growth performance of two Nile tilapia (Chitralada and Red-Stirling), their crosses and the Israeli tetra hybrid ND-56. *Aquac. Res.* 36, 1049–1055.

[74] Majumdar, K.C., Nasaruddin, K., Ravinder, K., 1997. Pink body colour in Tilapia shows single gene inheritance. *Aquac. Res.* 28, 581–589.

[75] Varadaraj, K., 1990. Dominant red colourmorphology used to detect paternal contamination in batches of Oreochromis mossambicus (Peters) gynogens. *Aquac. Res.* 21, 163–172.

In: Effects and Expectations of Tilapia ... ISBN: 978-1-63463-307-9
Editors: M. A. Liñan Cabello et al.

Chapter 3

TECHNOLOGICAL RECONVERSION PROPOSALS FOR TILAPIA: COLIMA, CASE STUDY

Marco A. Liñan Cabello*[*1], *Alfredo Hernández-Llamas*[2], *Laura Flores Ramírez*[1], *Carolina Sánchez Verdugo*[2], *Gerardo Verduzco Zapata*[1] *and Claudia Johana García-Olea*[3]

[1]Acuacultura/Biotecnología, FACIMAR, Universidad de Colima, Manzanillo-Barra de Navidad, Manzanillo, Colima, México;
[2] Centro de Investigaciones Biológicas del Noroeste (CIBNOR), Mar Bermejo Baja California Sur, México
[3]Secretaría de Desarrollo Rural, Gobierno del Estado de Colima, México

ABSTRACT

The region of Colima in Mexico currently development plan with some progress, however despite having increased the number of farms for growing tilapia, reported generally low yields as a result of organizational problems, water use, profitability culture, socioeconomic conflicts between the exercise of such aquaculture and agriculture. This chapter

[*] Email: linanmarco@hotmail.com.

discusses various alternatives with potential application to be considered for a program of technological reconversion both systems production extensive, semi-intensive and intensive

INTRODUCTION

In the last decade, aquaculture in Mexico has increased the number of farms and acres of production ponds. However the increase in total production has not been substantial. Much of the Aquaculture Production Units (APU) emerged without conducting the required previous studies, such as final design, approval of water use, environmental impact statement and staff training. Moreover, a considerable proportion lacks the necessary permits for aquaculture and scarcely has health certificates.

Figure 1. Location of the municipalities of Colima-México on the Pacific coast.

Currently in the state of Colima (Figure 1), there are about 100 APUs, where the predominant activity is tilapia and shrimp culture. In low proportion, the channel catfish farming is done in bicultural with tilapia as a species of greatest importance. Shrimp culture in inland waters notable for its importance with 1,384 tons in 2010 respect to 344 tons of harvested in tilapia same year. The statistics by municipality show that farms in Tecomán involved with 75% of the national production, ie 1,139 tonnes, followed by Colima and Manzanillo with 254 and 56 tons respectively. These high levels of production in Tecomán are mainly due to intensive white shrimp culture in inland waters as well as the largest number of hectares planted [1].

This has generated contrasts in cropping systems and between species grown in the state. Particularly in the state of Colima, the case of inland shrimp, generally lead APUs proper crop management, while tilapia, most of the units makes the crop under a scheme of low technical assistance, high water exchange, lack of physical-chemical and calculation of feed conversion monitoring and large variation in cultivation techniques. Regardless of the crop species, multiple organizational problems, water use, crop yield, conflicts of socioeconomic nature between the exercise of aquaculture and agriculture, among others are recognized.

Thus, no immediate action to implement measures and aquaculture producers of the entity were in a disadvantage relative to other producers. This leads to the search for alternatives in this regard, restructuring production (RP) is an instrument involves modifying the traditional pattern of production through a comprehensive transformation where technological changes involved the conversion of crops, the conversion production and recovery of degraded areas. The same RP as proposes four areas of support: research, technology transfer, training and technical assistance and organization of producers.

Analysis of Aptitude

The following section provides a brief analysis of aquaculture suitability of the municipalities in the region of Colima in order to characterize those environmental, vulnerable and operating conditions that may limit the cultivation of tilapia and thus propose and promote specific actions performed part as a plan of RP. Under this context, the proposal promotes the RP looking to tap the potential of the aptitude of the area, some technological changes are recommended in order to improve productivity, competitiveness and sustainability and in some cases conversion to other species is a recommended aquaculture potential.

Table 1. Optimal condition; 2.-Condition recommended; 3. Moderately recommended Condition; 4. Distantly recommended Condition; 5.-Condition completely unfavorable

		MUNICIPALITIES									
AREAS	FACTORS	ARM	COL	COM	COQ	CUA	IXT	MZO	MIN	TEC	VA
Environ-mental	Temperature	2	2	3	2.5	3	2	2	4	2	3
	Soil Type	3	1.5	2.5	2.5	2	2.5	4	2	1.5	2
	Pending Land	2.5	2	2	2	1.5	3	2.5	3	2	2
	Water Availability	2	2.5	3.5	2	2.5	2.5	2.5	3	2	3.5
	AVERAGE	**2.38**	**2**	**2.75**	**2.25**	**2.25**	**2.5**	**2.75**	**3**	**1.87**	**2.62**
Acces To Services	Access To Electrical Network	3	3	4.5	3	3	4	3.5	4	3	3.5
	Overland Routes Of Communication	3	1.5	3.5	3	2.5	4	3	4	3	2
	AVERAGE	**3**	**2.25**	**4**	**3**	**2.75**	**4**	**3.25**	**3.75**	**3**	**2.75**
Social Vulnerabil-ity	Conflict With Other Activities	4	4	5	4	5	3.5	3	3	5	5
	Degree of Marginalization.	2	1	2	2	1	3	1	3	2	1
	AVERAGE	**3**	**2.5**	**3.5**	**3**	**3**	**3.25**	**2**	**2.75**	**3.5**	**3**
Climate Vulnerabilty.	Proximity to Rivers	2	3	2	3	2	2	3	4	2	3.5
	Precipitation	2	2	2.5	3.5	2	2	3	4	3	2.5
	Invasive Species	2.5	2	2	2	2	2	2	2	4	2
	AVERAGE	2.16	2.33	2.16	2.83	2	2	2.66	3.33	3	2.66
	GENERAL VALUE	**2.68**	**2.18**	**3.25**	**2.62**	**2.56**	**3.06**	**2.68**	**3.12**	**2.56**	**2.75**

ARM = Armeria; COL = Colima; COM = Comala; COQ = Coquimatlán; CUA = Cuauhtémoc; IXT = Ixtlahuacán; MZO = Manzanillo; MIN = Minatitlan; TEC = Tecomán; VA = Villa de Álvarez.

The definition of variables for the analysis of aptitude to environmental requirements for growing tilapia, access to basic services and supplies, as well as the vulnerability of social and climatic. This analysis was conducted at the municipal level and consisted of a linear rating scale from 1 to 5, where 1 represents an optimum condition and 5, a totally unfavorable condition (Table 1).

Regarding the environmental factors evaluated, we distinguish that municipalities with best potential for growing tilapia are Armeria, Colima, Coquimatlán Ixtlahuacán, Manzanillo and Tecomán. Particularly, Cuauhtémoc, Comala and Minatitlan have temperatures not preference for the cultivation of bout species, so that the farms in these locations may have a higher calling for the cultivation of other species. Regarding water availability, Comala, Minatitlan and Villa de Alvarez presented the least acceptable values. Whereas current technology tilapia farming requires large volumes of water, such a situation could worsen in those municipalities.

With regard to basic services that could support technological change, there are municipalities with limited access, among these, Comala, Ixtlahuacán and Minatitlan. This would require major investments, primarily related to access to the electricity network and terrestrial communication pathways.

As for social vulnerability, particularly related to conflicts with other sectors for water use and the degree of marginalization, the municipalities of Comala and Tecomán Ixtlahuacán exhibit the worst values in the scale of recovery. This suggests greater attention from policy makers and the use of technologies that allow the rational use of water.

The climate vulnerability evaluated respect to the nearness of the APUs to rivers, precipitation levels and affectation around invasive species are higher in Coquimatlán, Manzanillo, Minatitlan and Tecomán. In this aspect, the meteorological events that cause major damage in the region are, first, floods, hurricanes and rains followed. The higher probability of affectation and poor implementation of preventive measures have been revealed in the evidence and damage historically recorded in the municipalities of Tecomán, Armeria and Manzanillo.

A recent factor that can lead to a problem of increasing significance in tilapia framing systems in particular, is the proliferation of invasive species, namely the devil fish and is increasing the populations of crocodiles. This latter has populated bodies of water Tecoman (Amela and Alcuzahue lagoons), which has a negative impact on the farming practices of tilapia in cages. Thus, failure to take appropriate action, in Tecomán aquaculture and the rest of the municipalities can be seriously affected. Also, this effect can also limit the choices of RP in tilapia farming and / or other species of interest.

Notwithstanding the foregoing, and in order to get a better perspective on a plan RP in tilapia culture is necessary to assess a more local scale whereas other indicators and mechanisms to translate technological plan recommendations into a productive reconversion.

Restructuring Policy- Inland Waters

The Lagoons Alcuzahue and Amela are the longest inland in the state of Colima, are located in the coastal plain of the valley of Tecomán and both are of great importance for the development of agriculture.

Amela Lagoon is located in the southeast of the state of Colima, between the coordinates 18 ° 48'05 "N and 103 ° 46 '34" W, within the municipality of Tecomán. The lagoon has a total area of 1,160 hectares, the area of the lake has mostly type A (Wo) warm subhumid climate with summer rains, and lower humidity. North of the lake a BSi (h ') very warm to semi-dry climate type is present, the average annual temperature is 25.2 ° C, and the average annual rainfall is 837.2 mm [2].

The next major body of water is the lagoon Alcuzahue which is also located in the Municipality of Tecomán, Colima. Located at coordinates 18°55'0" N and 103°46'59" W. It is located at a height of 30 meters above average sea level. This body of water is part of a strip of tulle in contrast to the wild vegetation of the mountains. The main species here caught are: tilapia, catfish. Access to the lake is by a dirt road in good condition. In this lagoon crocodile which represents a point of touristic interest and for the economy of this region.

In these water bodies, the most important fishery resource is the gray tilapia which for many years has represented the livehood of fishermen from 60 families belonging to three cooperatives. Also there have been some type aquacultural experiences among these, farming tilapia in cages. However, the fisheries and aquaculture production of this resource has decreased significantly due to environmental factors and anthropogenic factor, increasingly affecting the standard of living of the fishermen.

Currently, the main problems associated with low productivity aquaculture and fishing in this lake are:

- Increase in the population of the American crocodile (*Crocodylus acutus*) and its impact on traditional fishing
- Introduction and proliferation of devil fish (*Hypostomus Plecostomus*) and their effects on natural populations of tilapia (Figure 2).
- Proliferation of aquatic plants
- Lack of support to upgrade and / or update the fishing and aquaculture infrastructure.

Figure 2. Left: Specimens of devil fish. Right: Specimens collected in one haul of gear.

It is possible to distinguish that this problems related to low productivity is partly dependent on external factors such as the effect on water quality by agricultural practices, on the other hand, serious effects caused by the introduction of exotic species are detected, particularly devil fish, and alligator population increase. All these problems somehow affect the productive capacity of the tilapia as the main fishery resource. In the particular case of the devil fish, there are areas of the lagoon abundance competing very significantly with the population of tilapia, in turn causing the loss of fishing equipment and tools for fishermen.

The main result is, lower economic income both fishermen and fish farmers.

The creation of a management plan could be considered that includes a strategy for a technological reconversion for the use of the devil fish. The above is based on the successful experience in other water bodies around the world. This organism can not only be a great source of food protein, but has identified a biotechnological interest as a source of enzymes, and input of significant value for the manufacture of silage as a preserved protein source.

On the other hand the creation of a management unit for the American crocodile is proposed, which can translate into a tourist attraction. The participation of government authorities and institutions help to promote the application of good agricultural practices and land use, which would impact on improving the quality of water that allow more sustainable transformation plans and make sustainable farming projects and / or tilapia reseeding plans.

An organizational improvement is also necessary within which the true interest and commitment by each partner is defined with respect to taking steps aimed at reproductive process transformation. It is important define groups with interest in fishing and only a group that shares the expectation of diversification specifically for aquaculture projects. According to exploration carried out previously, only Amela lagoon exhibits peripheral extensions (approx. 30 Ha) with potential for aquaclulture. However the main expectation conversion is represented by the use of existing infrastructure aquaculture, where in the past years a large aquaculture unit operated. This infrastructure consists of 19 ponds that allow the controlled tilapia stocking towards reproduction, in parallel with the use of ponds that can accommodate the intensive farming of tilapia, also other species of greatest commercial interest. This same potential is present in the periphery of the Alcuzahue lagoon.

With respect to the problem of devil fish it considered highly relevant to propose a reconversion plan for integrated technological use of this species and its control. There must be a realization of a arrangement that would aim to make an approach with the corresponding authorities. Additionally required that municipal, state and federal authorities should promote the implementation of good practices in agriculture which would improve water quality, which in turn support other productive alternatives.

Productive Reconversion: Tilapia Case

Considering the various problems and priorities of the tilapia industry in the state of Colima it was possible to determine some alternatives for productive renewal, promoting the efficient use of water, energy, raw materials of the region, it aims at sustainable development of tilapia farming (Figure 3). As shown in Figure 3, the interaction of certain conditions form part of the success in tilapia culture systems: the implementation of an adequate water supply, with characteristics of temperature, oxygen and pH conducive, access to basic services and commodities such as balanced food, electricity, fingerlings machinery, equipment and protection against predators in all production stages; besides complying with certain rules and regulations and other tax provisions. All this, linked with the basic domain of culture techniques and the existence of previous production records, can provide the basis for the implementation of the principles referred to in the alternative or restructuring of production plan, based on technology transfer and in the execution of a marketing plan, the medium and long term, increase the yield, reduce the cost of food supplies and achieved the activation in the tilapia industry by obeying a technicized and sustainable plan (Figure 3).

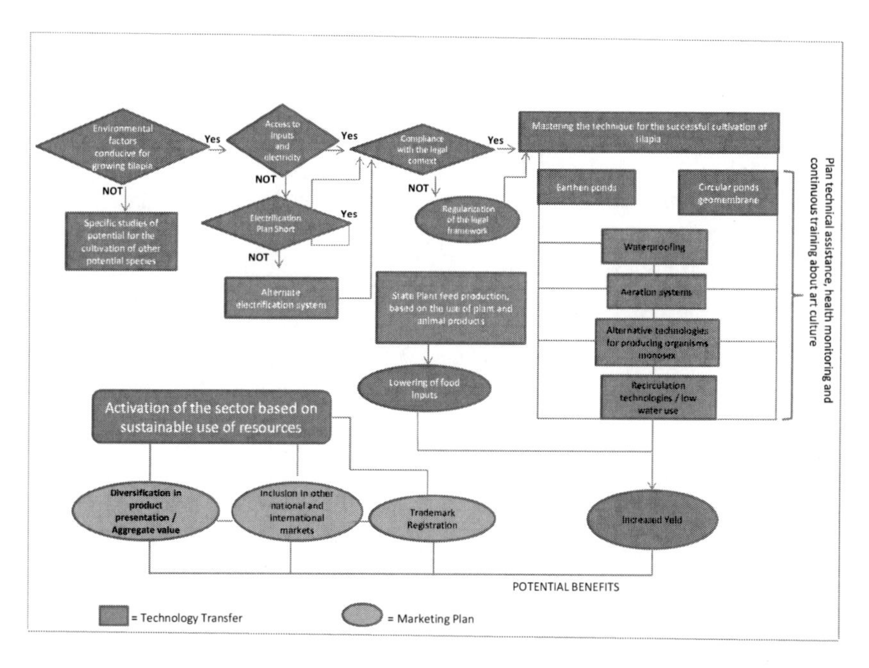

Figure 3. Plan of technological recommendations for productive reconversion in tilapia farming.

Under the same context, it may be noted that fish farms established in Armeria, Colima, Coquimatlán Ixtlahuacán, Manzanillo and Tecomán have a better aptitude for growing tilapia, given environmental conditions (Table 1). However, as noted above, in each municipality a more specific assessment is needed in order to verify the proximity or access to basic services and supplies, as necessary, or limiting, for the implementation of a plan was drawn reconversion.

ALTERNATIVE ELECTRIFICATION SYSTEMS

According to the analysis of eligibility criteria, most farms do not have such services, mainly because of the distance to the posts and wiring of electrical and / or the poor state of the terrestrial routes. In this regard, the implementation of renewable energy sources, in particular the use of solar energy could play an important role in overcoming this obstacle.

In that sense, the geographical location of the state of Colima is ideal for utilization of solar energy, because solar radiation is sufficient to provide the total energy demand a production unit. Therefore, the adoption of these technologies also would contribute to mitigation and adaptation to climate change, a key part of national policy, considered in the Special Climate Change Program.

Some alternatives of using solar energy to generate electricity are photovoltaic systems, among which the autonomous photovoltaic system (APS) and the interconnected photovoltaic system (IPS).

An APS is independent of the electrical system, and depends only on its components (solar panels) to generate electricity. They are used where conventional electricity is not available or is unreliable or when regular power supply for the cost of a remote location, require placing poles and wiring. They are commonly used for pumping water without generating pollution or noise. Operate day and night regardless of inclement weather.

On the other hand, IPS basically consists of a photovoltaic generator coupled to an inverter that operates in parallel with the conventional electricity grid. In this scheme, the quality of the electrical energy generated by the APS is similar to the conventional electricity grid. When there is a deficit between the demand for electricity in the house regarding the photovoltaic generation, this spread is covered with conventional electricity and vice versa. These systems are basically in urban and rural areas and are used primarily for lower costs in the electricity tariff.

Besides the above mentioned systems, there are other technologies that help reduce the problems: energy and environmental among which are, solar water heaters, biogas digesters, and biofuels, among others. Such systems may or may not be joined together, depending on demand and the sector involved.

Particularly, APUs established in the municipalities of Comala, Ixtlahuacán, Minatitlan emerge as candidates to use these alternative systems, given their apparent remoteness from the electrical system. However, state-level access to the electricity grid on farms is also a limiting factor in adoption of new technologies.

Currently, the biggest obstacle or barrier present for photovoltaic technology is the high initial investment compared to other technologies. However, in each country there are agencies which provide incentives to promote sustainable practices, such as harvesting, generation and use of renewable energy, energy efficiency and sustainable food production.

COMPLIANCE AND REGULATION OF THE LEGAL FRAMEWORK

This section only considers the legal regulation of the farms, as most of them operate outside the law. This does not require a technological recommendation, however, breach the rules, regulations and other tax provisions, excludes or limits of possible financing schemes of the state and federal government, further limiting the possibilities for technological reconversion.

WATERPROOFING EARTHEN PONDS

According to its texture, the soils are classified as coarse or fine. The fine ground fraction is composed of clay particles smaller than 0.074 mm, its size nearly indistinguishable (not visible) to the human eye. The ground coarse fraction (particles larger than 0,074mm) is constituted of sand and gravel. Thus, a clay soil and low permeability is therefore more favorable for pond construction. Sandy soils, or with lots of pebbles and stones, generally have high infiltration, demanding increase water use. These soils are also less stable and more susceptible to erosion.

In the state of Colima, areas with fine sediments found in the regions of Coahuayana river; in most of the municipalities of Colima and Tecomán. The coarse textured soils are located in the municipalities of Armory, Comala, and Ixtlahuacán Coquimatlán and partly Tecomán and Manzanillo as the municipality with the highest percentage of coarse soils. Because individual cases of filtration problems have been detected problems on some farms in the state that have rustic ponds that (Villa de Alvarez, Tecomán) proposed some mitigation strategies:

- Soil compaction and edges of the pond
- Application of organic fertilizers to clog soil pores
- The use of particulate dispersants (sodium chloride, sodium hydroxide)
- The use of clays with high swelling and water absorption (bentonite)
- Coating with blankets high density polyethylene (HDPE) and cloreto polyvinyl (PVC)
- Implementation of concrete.

It should be noted that the choice of each depends on the characteristics of the site and a detailed assessment of costs for implementation and operation. However, it must prioritize the rational use of water in order to avoid or minimize chronic or seasonal water supply problems and make continuous operation of APUs viable.

In Mexico some agencies have established programs that provide incentives to support investment in equipment and infrastructure, in concurrence with the states, for different strategic lines, including for infrastructure and equipment Aquaculture Unit, which includes support to counter or minimize water seepage problems since it is mainly coarse soils.

AERATION SYSTEMS

In general, tilapia farming systems in the state of Colima is operated without proper aeration. Meanwhile in aquaculture most problems including diseases are caused by poor water quality however, most of the problems of water quality can be solved with a proper aeration, it is clear that it plays a decisive role in this process.

The lack of an aeration system on farms is directly related to the lack of electricity, either by the distance to wiring network and / or because production costs rise significantly. Photovoltaic systems mentioned above stand out as an option. A good aeration system allows, among other things, increase crop densities, maintaining optimal growth and reduce stress on fish; this generates a high resistance to disease and better water quality.

In the state of Colima, tilapia farming technology is mainly in earthen ponds under extensive and semi-intensive systems, although there are a significant number of farms employing geomembrane tanks. In such systems the importance of maintaining adequate levels of oxygenation, is an aspect that is usually unknown or ignored.

Table 2 shows some ventilation systems and their ability to transfer oxygen. The ventilation system for a pond requires knowledge of certain aspects like the mirror of water, the depth of the pond, the duration the system must operate, the oxygen needs of the crop species, among the most significant.

Table 2. Types of aerators and aeration efficiency

Type of aerator	Average oxygen transfer (kg.O_2/kW-hour)
Paddle Aerator	2.13
Injection – Mixer pumps	1.58
Vertical pumps	1.28
Dispensing pumps	1.28
Air Diffusion System	0.97

Alternative Technologies for Producing Organisms Monosex

The biggest problem in the cultivation of tilapia is excessive reproduction, causing overpopulation of the pond and consequently reduced growth of fish. To prevent this problem, the ponds should be planted with only male specimens. This technique is called monosex tilapia and is used when needed to produce large fish for the market. Multiple efforts have been made by researchers to try reduce the effect of early breeding tilapia. Among the techniques applied manual sexing, hybridization and sex reversal, but the latter is the most used in commercial-scale production.

As in other regions of the world where tilapia is grown, the main problems that need to be resolved are:

- Technical deficiencies in the production of fingerlings which increases inbreeding factors which adversely affects the yield and quality of tilapia.
- The increase in the price of nutritious food for fish feeding causes an increase in production costs which affects the profitability of the activity.

- For the production of monosex hatchlings hormonal compounds are used in some countries which represent a risk factor for human health. This limits the expectations of exporting to other countries, particularly the United States as one of the main buyers of tilapia in the world.
- Some farms grow males and females together, inducing an "early reproduction" causing the presence of small fish with no commercial value.
- Overall gray tilapia is frequently associated with the taste and smell of mud, which subtracts from market acceptance.
- *O. niloticus* or Nile tilapia, of greater commercial exploitation is limited to tilapia culture in freshwater.
- There is no genetic diversity of red tilapia groups that are adaptable to different environmental conditions on the farms.
- The red tilapia have low rates of growth, with production cycles 6-9 to reach commercial size.
- Cultured red tilapia's low percentage of muscle mass.

Pargo-UNAM One of the most cultivated species in the world is the of Nile tilapia wild type or gray mainly due to its rapid growth, there are another group called red tilapia for their attractive coloration are demanded by a large segment of tilapia market; however, usually red tilapias have a lower rate of growth relative to the gray Nile tilapia, so that a red tilapia farmer invests an additional 2 to 4 months to get a red tilapia commercial weight (400g).

The Pargo-UNAM red tilapia consists of genetic groups: Rocky Mountain (25%), *Oreochromis niloticus* (25%) and Florida Red Tilapia (50%). The advantages of Pargo-UNAM are similar to the gray Nile tilapia growth at the end of the crop a weight 40 to 60% higher than in other red tilapia and can be grown both in freshwater, brackish and saltwater environments.

Currently, the supply of Pargo-UNAM is run by The Center for Teaching, Research and Extension in Tropical Livestock (CEIEGT), in the state of Veracruz, Mex. [3]. However, CEIEGT still requires investing in the infrastructure to produce large-scale tilapia meet domestic demand.

Genetically Male Tilapia (GMT) and / or Super males-This technology, called the "YY male technology" arises out of a breeding program that combines feminization and progeny testing to produce a novel type of males with YY genotypes (that is, with two male sex chromosomes) instead of the usual XY males genotype. These males are called YY "supermale" in these specimens, no genetic engineering or genetic modification is used, it is simply considered a first phase of feminization (hormonal) and a second phase of progeny testing, these fertile and produce viable male only male. These individuals are GMT excellent results in survival, decreased territoriality, high feed conversion efficiency, minimal variation in size, high growth rate, high yields by weight and decreased harvest time.

Although advances in the genetic improvement of tilapia are encouraging, a project that should be considered is the creation of a center producing monosex hatchlings from GMT, as a strategy to promote the aquaculture industry through sustainable technologies to get set priority of the mix program federal funds. This would represent an additional advantage, considering they would not be with the final product, retail market, hormonally treated specimens, which at this time, is a barrier to the appreciation of the product and especially so in the future be allowed to expand the marketability of export quality.

Specimens produced via GMT could be considered as a strategic element for the appropriate authorities to be proposed as a priority in the program of support to the aquaculture sector.

RECIRCULATION / LOW WATER CONSUMPTION TECHNOLOGIES

In the region of Colima, water scarcity is an obvious problem; some municipalities already show problems with it´s availability and quality. This has generated interest in knowing the current status of the availability and quality water. In this regard, Pastén-Zapata et al. [4], indicate that in the state of Colima quality of surface water is moderate to low, where the Coahuayana River has lower quality indices followed the Marabasco river. Regarding the availability of water in the sub-basins of the state (Marabasco, Armeria and Coahuayana) found that the degree of pressure on the resource is moderate to strong. The Armería sub basin has a stronger degree of pressure with respect to the other.

In relation to water from underground sources, the quality is acceptable. However, in some aquifers located in the municipalities of Armería, Manzanillo and Ixtlahuacán, limits referenced in the Mexican Official Standards potable water are exceeded. Moreover, in this same study it mentions that the aquifers are vulnerable because groundwater levels near the Earth's ground surface layer also exhibit high permeability.

A possible strategy to improve the situation and resource demand is the comprehensive management of the resource; consider actions that optimize the different uses of water in the sectors involved. In this regard, traditional aquaculture requires large amounts of water and large tracts of land, therefore, recirculation systems (RS) or closed systems are presented as an advantage in the rational use of this resource.

In these systems, the amount of exchange is less than 10% of the total daily volume of the system. In addition, it is possible to better control the monitoring and control of physical chemical parameters such as temperature, salinity, dissolved oxygen, carbon dioxide, hydrogen potential (pH), alkalinity and nitrogen metabolites such as ammonia, nitrites and nitrates. This allows continuous production throughout the year and in the farmed organisms, may represent better growth rates and feed conversion (Table 3).

Table 3. Advantages and disadvantages in using recirculating aquaculture systems

ADVANTAGES	DISADVANTAGES
Reduce the transmission and spread of diseases. High initial investment cost	High initial investment cost
Decreases considerably polluting the environment	Qualified personnel are required
Optimizing the use of resources such as water, food, energy, land, personnel, etc.	
Higher levels of feed conversion factor	
Production plan more efficient	
It is possible use at different stages in aquatic organisms both freshwater and seawater	

For a recirculation system to be efficient and provides a suitable environment it must possess five main attributes [5]:

- Removal of solids
- Biofiltration
- Aeration
- Degassing
- Water circulation

Given the high initial investment in the implementation of a RS, as the cost of importing the technology from USA is very expensive, the challenge both in our region of study is to develop these technologies using materials that are readily available in the region, so that the production capacity per unit of capital invested is maximized. That is, recirculation systems developed inexpensively and efficient according to the needs. Another important aspect of this technological innovation, is the need to develop constant research, and might represent an opportunity for the cultivation of other species or native aquaculture potential in a, sustainable and friendly way with the environment. This paradigm can be one of the investigations to be undertaken by a research and technology transfer in the state of Colima.

PLANT GOVERNMENT OF FOOD PRODUCTION

In tilapia farming, as in any other species, the most expensive input cost is food which can represent up to 60% of the production cost and involve a strong and stable investment. In addition, some of the foods manufactured by enterprises in the sector do not meet the actual components of energy, protein, vitamins, fiber and minerals, necessary for the proper development of the fish. Therefore, one of the main requirements of the activity is to establish a feed mill products and subproducts that are available in the region in search of promising high quality materials, high nutritional value and low cost in the region, which in turn, enable the manufacture of mixtures of quality; acting directly on reducing production costs of tilapia.

In this regard, federal and state agencies developed a project called "Technology Transfer for the development of balanced feed for tilapia with inputs from the region". To this end the state of Colima mapping to determine the quality and chemical composition of the main ingredients used in feed for tilapia, sampling grasslands, crops, ingredients and by-products using traditional and non-traditional. Currently already has this characterization, so it is possible to formulate food rations to help with the daily requirements of tilapia. According to a Mexican research institution of forest, agricultural and livestock research (INIFAP), recently made available to producers a database that includes the characterization of 56 products and by-products it also includes a spreadsheet to facilitate the formulation proportions for higher nutritional value.

However, to attain the objective of establishing a food processing plant it is necessary to cover other aspects such as the acquisition of machinery, equipment and tools, and among others, making it necessary to identify funding sources. Currently there are foreign companies offering technology for feed production aimed at satisfying local demands with production capacity of 2.5 ton / day, which could be suitable for the tilapia industry in Colima. Implementation of these technologies do not represent large investments, and if considerable benefits for the use of agro-industrial products in the region that would impact in lowering dietary intake.

Conclusion

Currently the state of Colima, similarly the other states of Mexico, tilapia farming is just beginning, however adaptive and productive capacities of this species can be of great importance to increase production levels, provided they consider adoption of new technologies aimed at sustainable use of resources. The lack of technological capability of adaptation can be a great risk factor that would affect the productive capacity and competitiveness of farmers

Acknowledgments

This project was supported by the Fondo Mixto Conacyt-Gobierno del Estado de Colima" (COL-2011-C03-186334). The authors acknowledge of the facilities granted by Adalberto Zamarroni Cisneros, Secretary of Rural Development of the Government of the State of Colima.

References

[1] Fracchia-Durán, A. and Liñán Cabello, M.A.. Diversificación y desarrollo sustentable de la acuicultura en Colima. Uso de un modelo multicriterio para la diversificación y manejo sostenible de la acuicultura: Caso de Estudio. Editorial Académica Española, Alemania. 357 pp. 2013.

[2] Arredondo-Vargas, E., Osuna-Paredes, C. and De Jesús-Avendaño, C. Opinión Técnica para determinar el establecimiento de un periodo de veda de tilapia en la Laguna de Amela, Colima. Instituto Nacional de Pesca, CRIP Pátzcuaro. 8 pp. 2012.

[3] Calixto Escobar and Nicte-Ha. Pargo-UNAM: Una alternativa en el mundo de la acuicultura tropical / Nicte-Ha Calixto Escobar. México: IICA, 2011.

[4] Pasten-Zapata, E., Calderón Zúñiga, H., Mahlknecht, J., López-Zavala, M.A., Caballero García, C.A. and Horst, A. Evaluación diagnóstica de la calidad y disponibilidad del agua superficial y subterránea en el estado de Colima. En: Laclette, J.P., Zuñiga, P. & Romero, J.A. (eds). El impacto de los Fondos Mixtos en el desarrollo regional, Vol. I. CONACyT, Foro Consultivo, Científico y Tecnológico, REDNACECYT. 408 pp. 2011

[5] Summerfelt, S., Wilton, G., Roberts, D. and Savage, T., Developments in Recirculating Systems for Arctic Char Culture in North America. Proceedings of the Fourth International Conference on Recirculating Aquaculture, Roanoke, Virginia. 2002.

In: Effects and Expectations of Tilapia … ISBN: 978-1-63463-307-9
Editors: M. A. Liñan Cabello et al.

Chapter 4

PRESENCE OF TILAPIA IN CUYUTLAN LAGOON COLIMA, MÉXICO

M. Patiño-Barragán[1,*], M. A. Galicia-Pérez[1], S. Quijano-Scheggia[1], C. Lezama-Cervantes[2], M. A. Liñán Cabello[2] and A. Morales-Blake[2]

[1]Centro Universitario de Investigaciones Oceanológicas, Universidad de Colima, Manzanillo, Colima, México
[2]Facultad de Ciencias Marinas, Universidad de Colima, Manzanillo, Colima, México

ABSTRACT

Misconduct by aquaculture management in the state led to the introduction of tilapia in the Cuyutlán Lagoon in Colima, Mexico. To determine their distribution and abundance, 20 sampling stations were implemented in the four vessels of the lagoon system for a period of five years (2008–2011). Catches were realized monthly using the traditional technique of fishermen in the region and a net of 2.0 in mesh, 1.5 m wide and 100 m long. During the study period, 1,600 tilapias were collected with a biomass of 162.6 kg; standard length and weight averages were 13.8 ±8.2 cm and 91.3 ±13.2 g, respectively. Two species were identified: *Oreochromis niloticus* (Linnaeus, 1758) and *Oreochromis mossambicus* (Peters, 1852), with a presence across the lagoon that increased during the

* E-mail: mpkile@ucol.mx.

rainy season. Vessels III and IV were the most abundant, especially in shallow areas barely influenced by sea communication. Comparatively, abundance was 44.8% for *O. niloticus* and 55.20% for *O. mossambicus*. Males were more abundant than females in both species. The influence of seawater by opening Tepalcates Channel has reduced its presence of this specimen in the Vessels I and II. Both species are adaptable and can pass from continental freshwater environments to marine coastal, particularly to the hypersaline conditions of Vessel IV, which shows drastic changes in environmental parameters. Here, *O. mossambicus* stands out with an abundance of 78.3% compared to *O. niloticus*. Successful acclimatization of both species suggests the presence of a genetically selected generation, with possible consequences to the lagoon's trophic structure.

INTRODUCTION

Interest in the study of exotic species is growing worldwide due to their effects on native communities and global biodiversity [33]. Until 2004, exotic fish species in Mexico numbered 118 [36]; aquaculture introduced 38 of these [3].

Exotic fish are responsible for a great deal of biological, ecological, economic, and social damage to Mexico and have caused native species to disappear in more than 100 scattered localities across the country. They have negatively affected important fisheries, generating economic hardship in certain areas [1].

The global aquaculture of tilapia is on the rise; it is the second largest group of farmed fish after carp. In 2005, world production of tilapia equaled 2,692,594 tons. The Asian continent was the largest producer with 63.2% of the world total, followed by Africa and America with 26.6% and 10.0%, respectively [14]. After years of delay and with the hope of improving the diet and living conditions of rural communities, the tilapia was officially introduced to México in1964 [12]. In the state of Colima, tilapia farming began in 1973 [29]. Thereafter cultivation increased slowly, producing only 130 tons per year in 2010 [35]. However, despite this and due to poor aquaculture management, tilapia have infiltrated Colima's Cuyutlán Lagoon. This mismanagement has produced another problem of equal or greater magnitude than the first. In the 40 years since its introduction to the state as a crop species, tilapia have colonized many fresh, brackish, and marine water bodies in the Colima territory.

Tilapia are aggressive and tolerant to wide variations in salinity, temperature and dissolved oxygen concentrations. They eat a wide variety of food sources, adapt well to different environments, and demonstrate phenotypic plasticity and high reproductive efficiency—all characteristics that promote colonization. The Global Invasive Species Programme (GISP) rated tilapia as one of the 100 most dangerous invasive exotic species [39]. The purpose of this research is to analyze their presence in the Cuyutlán Lagoon from 2008 to 2012, a period in which Tepalcates Channel, an artificial marine intercom channel that opens to the sea, presented three scenarios, 1) clogged; 2) under construction (to extend it from 250 to 450 m wide); and 3) in operation; with variations in depth, area, and replacement of lagoon water.

MATERIAL AND METHODS

Twenty representative stations of physiographic heterogeneity were located in the Cuyutlán Lagoon and tilapia were caught monthly during the period from 2008 to 2012 (Figure 1). In order to monitor shallow sites during low or medium tide, the capture date was selected under conditions of high tide using the Chart Calendar Tide Prediction Center for Scientific Research and Higher Education of Ensenada, Baja, California (CICESE).

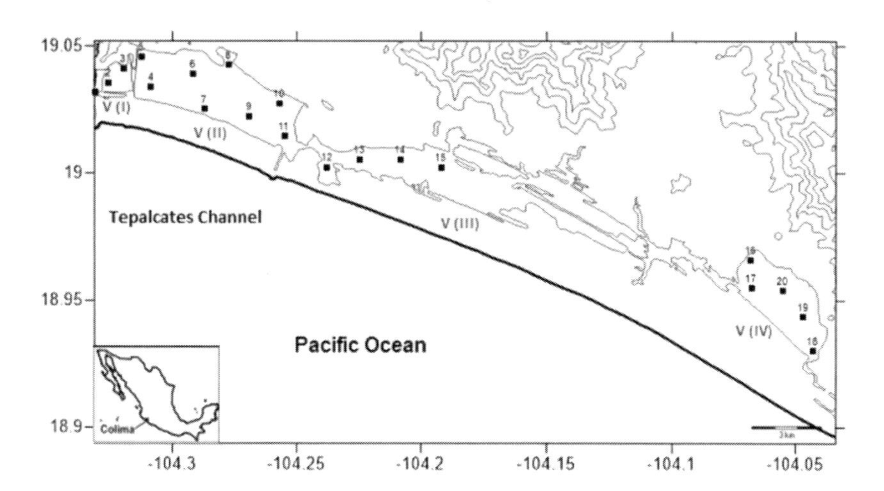

Figure 1. Locations of the 20 representative stations of physiographic differences in the Cuyutlán Lagoon (Vessels I, II, III, and IV) where monthly catches of tilapia were conducted from 2008 to 2012.

Fish were collected using an outboard motorboat equipped with a Garmin 72-H global positioning system (GPS) to locate sampling stations. Fish were captured at each station with a nylon 2.0 in mesh net, 1.5 m wide, 100 m long, using the regional fishermen's traditional technique of working the network in a circle for 15 minutes.

After collection, the tilapia were deposited into polyethylene bags and placed in a cooler at 8 ± 0.7ºC for transportation to the laboratory. The tilapia were then identified to species level using the taxonomic keys of Trewavas [12], Arredondo-Figueroa and Guzman-Arroyo [22], and Arredondo-Figueroa and Tejeda-Salinas [23]. Their standard length was measured and they were sexed and weighed.

Results

During the study period, 1,600 tilapia were collected in the four vessels of Cuyutlán Lagoon with a biomass of 162.6 kg and a standard length and mean weight of 13.8 ± 8.2 cm and 91.3 ± 13.2 g, respectively. Two species were identified: *Oreochromis niloticus* (Linnaeus, 1758) and *Oreochromis mossambicus* (Peters, 1852), with an abundance of 44.8% and 55.20%, respectively (Figures 2 and 3).

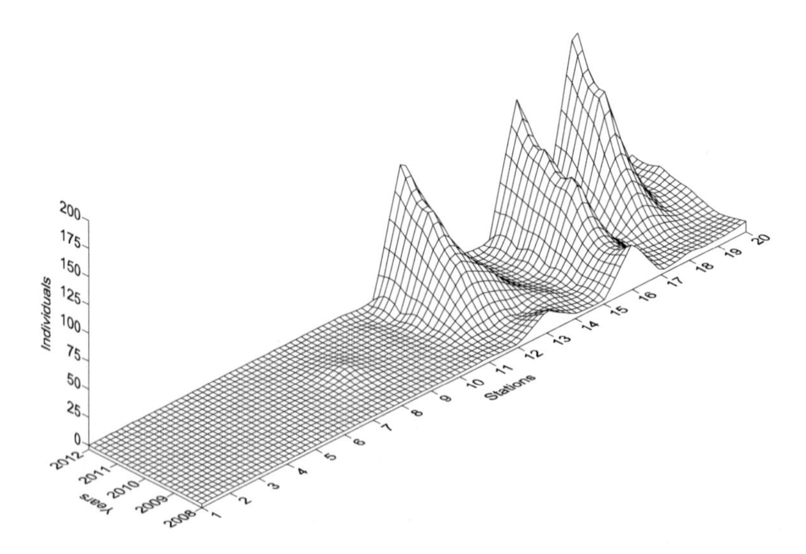

Figure 2. Abundance of *O. mossambicus* (individuals) from 2008 to 2012 in the Cuyutlán Lagoon.

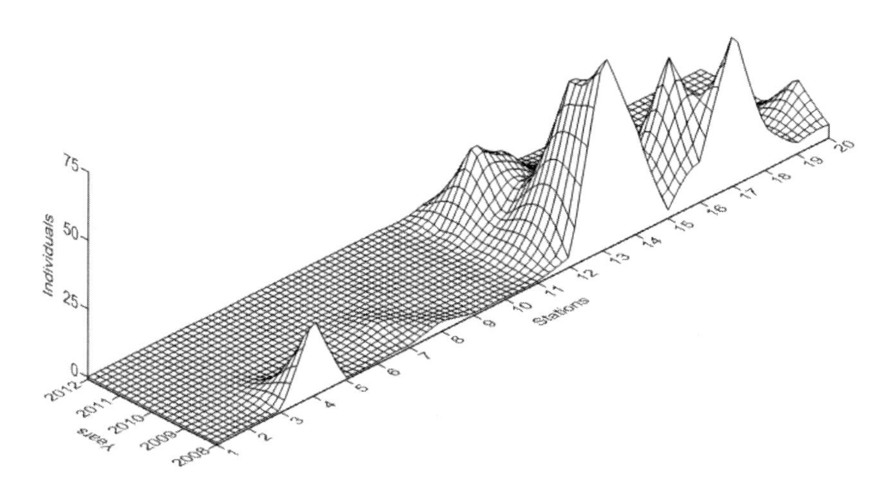

Figure 3. Abundance of *O. niloticus* (individuals) from 2008 to 2012 in the Cuyutlán Lagoon.

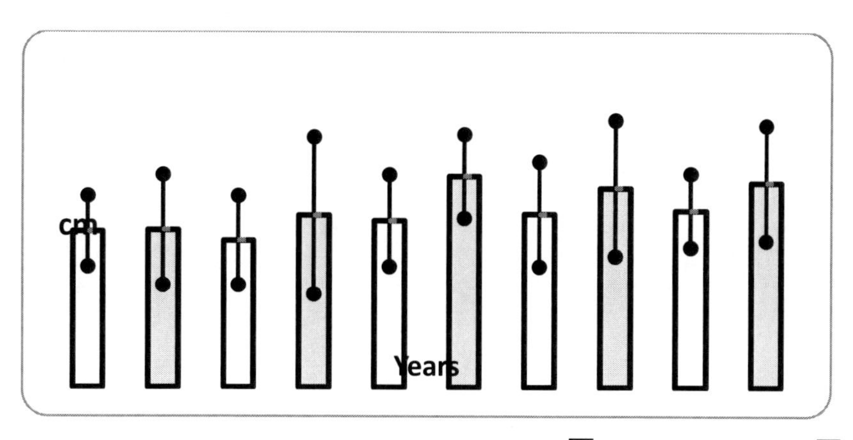

Figure 4. Annual average standard length of *O. niloticu s* ☐ and *O. mossambicus* ☐ caught from 2008 to 2012 in the Cuyutlán Lagoon.

The population of *O. mossambicus* increased from 2008 to 2012 while *O. niloticus* decreased. Both populations increased during the rainy season.

Average length and weight during the study period was 12.5 cm and 68.4 g and 14.4 cm and 108.0 g for *O. niloticus* and *O. mossambicus*, respectively, in the analysis of variance (ANOVA), however, no significant differences were determined ($P > 0.05$) in size and weight for *O. niloticus* or *O. mossambicus* for the years analyzed (Figures 4 and 5).

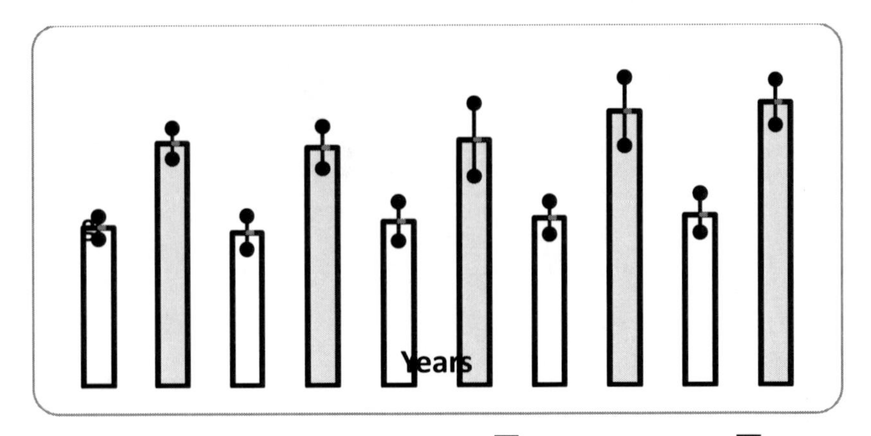

Figure 5. Annual average weight of *O. niloticus* ☐ and *O. mossambicus* ☐ caught from 2008 to 2012 in the Cuyutlán Lagoon.

Males were more abundant than females in both species: 57.7% and 52% for *O. niloticus* and *O. mossambicus*, respectively (Figure 6).

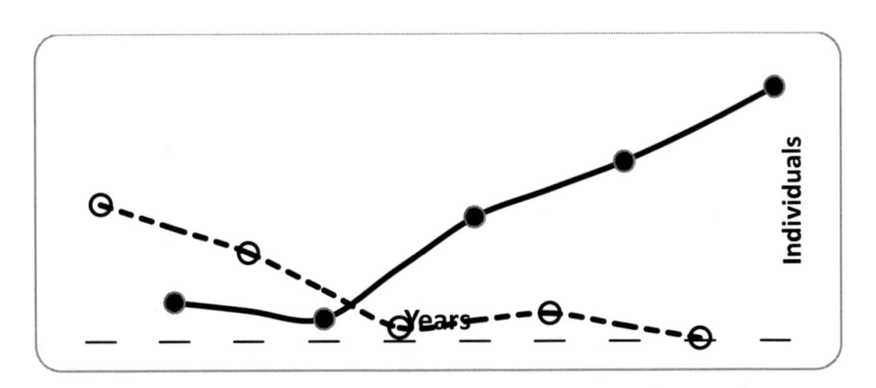

Figure 6. Proportion of males ☐ and females ☐ of *O. niloticus* -○- and *O. mossambicus* —●— captured from 2008 to 2012 in the Cuyutlán Lagoon.

The sites were more abundant in Vessels III and IV (seasons 12–20) in shallow areas where the influence of sea communications is very low (Figures 7 and 8). Opening Tepalcates Canal increased the presence of seawater in Vessels I and II. *O. niloticus* was more numerous in Vessel III, while *O. mossambicus* was more numerous in Vessel IV.

Tilapia are an invasive species in the Cuyutlán Lagoon and have been accidentally or intentionally introduced outside their natural range, showing the ability to colonize, invade, and persist (World Conservation Union) (IUCN) [1].

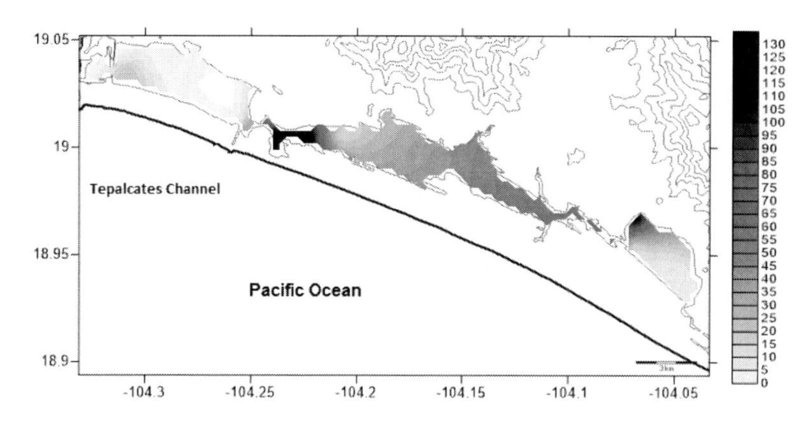

Figure 7. Distribution of *O. niloticus* (number of individuals) in the Cuyutlán Lagoon from 2008 to 2012.

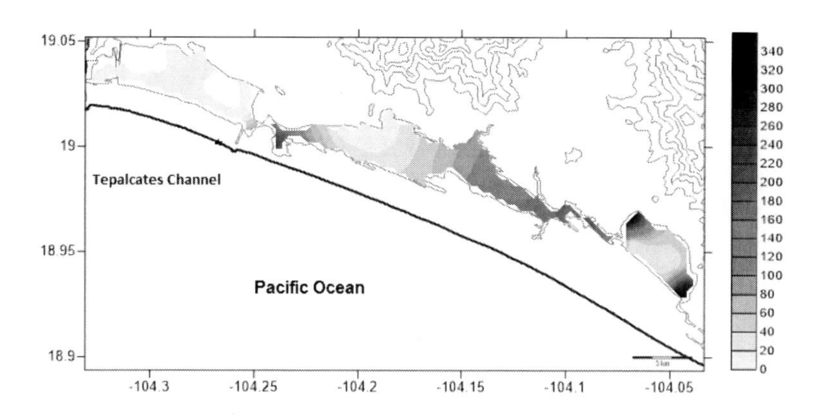

Figure 8. Distribution of *O. mossambicus* (number of individuals) in the Cuyutlán Lagoon from 2008 to 2012.

In a literature review of studies on ichthyofauna in the Cuyutlán Lagoon, Gómez [15] and Núñez-Fernández [11] found no mention of their presence; the first record of tilapia there dates to 1999 [8].

Based on this estimate, tilapia have been present in the lagoon since the early 1990s and have inhabited the lacustrine vessel for approximately 25 years as a consequence of aquaculture. During this period, despite the interannual variations of salinity (20–55 UPS), tilapia have been able to cover the three phases recognized for invasion [1]: introduction, establishment, and growth, as their presence was recorded during 2008 and 2009 in the lagoon, although in 2010–2012 they were concentrated in Vessels III and IV.

Tilapia genus *Oreochromis* in particular can survive direct transfer from fresh to saltwater [2] as a result of an adaptive mechanism that can regulate osmolality, the concentration of sodium ions and chloride [16]; [30], levels of cortisol, growth hormone [42], prolactin (tPRL177 and tPRL188) [18] in the plasma, and morphological changes in mitochondria rich cells (CRM) from the gills [31], which is crucial to the recruitment and regulation of NaCl and the acid-base balance of teleost fish [41] [24]. When the larvae of *O. mossambicus* are transferred from fresh water to seawater and vice versa, they are able to efficiently regulate the rate of water intake in a short time [26] and increase the apical surface area of mitochondria rich cells [27]. Although the average survival of fry in seawater is relatively lower than in fresh water [43], individual genetic differences suggest the possibility of selection for increased reproductive capacity of stocks.

The survival of both species of tilapia (*O. niloticus* and *O. mossambicus*) in brackish and marine waters of the Cuyutlán Lagoon is not unique; many examples demonstrate the proliferation of tilapia in seawater. In south Florida, USA, three species inhabit coastal habitats. Blue tilapia (*O. aureus*) has had the greatest impact, while Mozambique tilapia (*O. mossambicus*) is present in many localities and is common in the coastal waterways of southeast Florida and Tampa Bay. Black tilapia (*S. melanotheron*) was the first to be established in marine environments in Florida [5].

O. mossambicus also colonizes some Pacific islands: estuaries of Papua, New Guinea [20]; the brackish water of Tongatapu, Tonga, and Tuvalu [25]; the mangroves of Yap in the Caroline Island [38]; on Fanning Atoll in the Line Islands [32], and on the east and west coasts of Australia [6]. *O. mossambicus* and *S. melanotheron* have also been found off the coast of Hawaii [19] and have proliferated along Cuba's coast after release from reservoirs [21]. In Venezuela, adult tilapia with fingerlings have been captured in the mouth of the sea 100 meters from shore, suggesting that they have adapted to and are breeding in the marine environment [28].

A 9-year study (1999 to 2007) in the Cuyutlán Lagoon determined an average weight and standard length of 12.86 cm and 91.07 g for *O. mossambicus* [9] (Tepalcates was closed some years and open others). These findings are very similar to those obtained in the present work: 12.5 cm and 68.4 g and 14.4 cm and 108.0 g for *O. niloticus* and *O. mossambicus,* respectively. It is possible the small size and weight may result from predation or physicochemical changes.

A notable feature of the Cuyutlán Lagoon is its interannual variability in depth and salinity [10]. Both tilapia species' successful acclimation to these variations suggests the presence of genetically selected generations. Adaptability, especially of *O. mossambicus,* with an abundance of 78.3% compared to *O. niloticus* in Vessel IV, is manifest, allowing them to pass from a continental freshwater environment to a hypersaline coastal environment with drastic changes in environmental parameters during periods of rain and drought. Biological traits such as tolerance to broad ecological conditions, simple dietary needs, and rapid breeding with maternal care led to this species' popularity as a farm fish and produced an invader "model" [6].

Cuyutlán Lagoon has a history of manmade changes dating back many years [40], and one of these, Tepalcates Canal, has significantly affected the lagoon due to its communication with the sea. After the canal closed in 2008, *O. niloticus* was recorded in four vessels of the lagoon system. The presence of *O. mossambicus* was detected only in Vessels II, III, and IV. Between 2009 and 2011, construction to open the canal began, which affected both species. *O. niloticus* significantly reduced its presence in Vessels I, II, and IV; while *O. mossambicus* was minimally present in Vessel II, with increases in Vessels III and IV. In 2012, while the canal was open, the *O. niloticus* population was restricted to Vessel III, whereas *O. mossambicus* showed the highest increase in Vessels III and IV. It can be concluded that during this period, *O. niloticus* decreased while *O. mossambicus* increased.

Tilapia's presence in the lagoon demonstrates that the necessary precautions to prevent their proliferation have not been taken. Government entities remain ignorant of the impact of introduced species because they lack the warning systems and trained personnel to perform risk assessments.

The introduction of aquatic exotic species has been identified as one of the most critical environmental risks currently facing animal species, aquatic habitats, and biodiversity in general [4]. It has been associated with extinction in 54% of cases of native aquatic fauna worldwide [17], 70% of the fishes of North America [7], and 60% of Mexican fish [37]. Exotic species can affect native species through different mechanisms, including hybridization, competition for food and space, predation, disease transfer, alteration of habitat for native species, displacement of native species, alteration of the structure of the trophic levels, and introduction of parasites and disease [34].

CONCLUSION

Despite the very low aquaculture production of tilapia in Colima, poor methods of problem solving have generated another issue of equal or greater magnitude than the first. Forty years after its introduction to the state as a crop species, tilapia have colonized Cuyutlán Lagoon, which represents 92% of Colima's lagoon systems. Successful acclimatization of both species suggests the presence of a genetically selected generation, with possible consequences for the lagoon's trophic structure.

REFERENCES

[1] Aguirre Muñoz and R. Mendoza Alfaro, CONABIO, 2009 (México) pp 277-318.
[2] Fontaínhas-Fernandes, F. Russell-Pinto, E. Ma. Gomes, A. Reis-Henriques and J. Coimbra. *Fish Physiol. Biochem.,* 23, 307 (2001).
[3] Contreras-Arquieta and S. Contreras-Balderas, Vectors, biology, and impacts. Lewis Publishers (Boca Ratón, Florida USA) 1999 pp. 151-160.
[4] C. E. Hopkins, Actual and potential effects of introduced marine organisms in Norwegian waters, including Svalbard. Research report. Directorate for Nature Management, 2001 (Oslo Norwegian) 54 pp.
[5] E. Roberts, Non-indigenous fishes and invertebrates of concern to Florida's Gulf Coast. Paper presented at "Introduction of Nonindigenous Species Workshop", Gulf of Mexico Program. Metairie, 11 June, 1997 (Louisiana, USA).
[6] J. Russell, P. A. Thuesen, and F. E. Thomson, *Reviews in Fish Biology and Fisheries,* 22, 533 (2012).
[7] R. Lassuy, *American Fisheries Society Symposium,* 15, 391 (1995).
[8] Cabral-Solis and E. Espino-Barr, *Oceanides*, 19, 19 (2004).
[9] E. G., Cabral Solis, 2011. Centro de Investigaciones Biológicas del Noroeste S. C. Tesis de Doctorado. 2011 (La Paz, Baja California Sur, México) 134 p.
[10] E. Mellink and M. Riojas-López, *Revista Geográfica,* 142, 131 (2007).
[11] E. Núñez-Fernández, Tesis Doctoral, 1982. UNAM. (México, D. F.), 241 pp.
[12] E. Trewavas, *British Mus. Nat. Hist.*, 1983. (Londres, Inglaterra).

[13] J. L. Arredondo and G. S. Lozano, Universidad Autónoma Metropolitana, Secretaría del Medio Ambiente, Recursos Naturales y Pesca. Primer Curso Internacional de Producción de Tilapia. 20 al 22 de junio, 1996.
[14] FAO Estadísticas de Pesca y Acuicultura. FAO Yearbook, Anuario 2006 81 p.
[15] Gómez Cordero, Tesis de licenciatura, Universidad Autónoma de Guadalajara, 1974 (Guadalajara, Jal. México), 79 p.
[16] Assem and W. Hanke, *Comp. Biochem. Physiol.,* 64A, 17 (1979).
[17] J. Harrison and M. L. J. Stiassny, Causes, contexts, and consequences. Kluwer Academic/Plenum Publishers, 1999 (Nueva York, USA) pp. 271-332.
[18] D. Morgan, T. Sakamoto, E. G., Grau and G. K. Iwama, *Comp. Biochem. Physiol.,* 117A, 391 (1997).
[19] J. E. Randall, *Bull. Mar. Sci.*, 41, 490 (1987).
[20] J. Glucksman, G. West and T. M. Berra, *Biol. Conserv.*, 9, 37 (1976).
[21] J. Tucker and D. Jory, *World Aquaculture,* 22, 10 (1991).
[22] J. L. Arredondo-Figueroa and M. Guzmán-Arroyo, *An. Inst. Biol. UNAM. Ser. Zool.*, 56, 555 (1986).
[23] J. L. Arredondo-Figueroa and M. Tejeda-Salinas, *An. Inst. Biol. UNAM Ser. Zool.*, 16, 58 (1989).
[24] K., Shiraishi, T. Kaneko, S., Hasegawa and T. Hirano, *Cell Tiss. Res.*, 288, 583 (1997).
[25] R. Uwate, P. Kunatuba, B. Raobati and C. Tenakanai, A review of aquaculture activities in the Pacific islands region. Pacific Islands Development Program. 1984 East-West Center Honolulu, 441 pp.
[26] L. Y. Linn, C. F. Weng, and P. P. Hwang, *Biochem. Zool.*, 74, 171 (2001).
[27] Y. Linn and P. P. Hwang, *Physiol. Biochem. Zool.*, 74, 469 (2001).
[28] Nirchio and J. E. Pérez, *INCI,* 27, 1, 2001 at <http://www.scielo.org.ve/scielo.php?script=sci_arttext&pid=S037818442002000100007&lng=es&nrm=iso>. ISSN 0378-1844
[29] M. Patiño Barragán, A. O. Meyer-Willerer, M. A. Liñán Cabello, A. Mena Herrera, and C. Lezama Cervantes, *Iridia*, 5, 98 (2008).
[30] P. Hwang, C. M. Sun, and S. M. Wu, *Mar. Biol.*, 100, 295 (1989).
[31] P. Hwang, *Mar. Biol.*, 94, 643 (1987).
[32] S. Lobel, *Micronesica,* 16, 349 (1980).
[33] Bañón Díaz, *Revista Galega dos Recursos Mariños,* 3, 1 (2012).

[34] Bhaskar, and J. Pederson, 2003 at http://massbay.mit.edu/resources/pdf/ factsheet.pdf

[35] SAGARPA, Delegación Estatal Colima. Boletín junio del 2011, http:// www.sagarpa.gob.mx/Delegaciones/colima/Boletines/Paginas/2011B04 1.aspx

[36] Contreras-Balderas, Bases de datos SNIB-CONABIO, proyecto AE002, 2008 México, D. F.

[37] S. Contreras-Balderas, Vectors, biology, and impacts. Lewis Publishers, (Boca Ratón, Florida USA) 1999 pp. 31-52.

[38] S. G. Nelson, Marine Resources Management Division, Yap State Department of Resources and Development, 1987 (Colonia, Yap) 14 pp.

[39] S. Lowe, M. Browne, S. Boudjelas and M. de Poorter, 2004 UICN at //www.issg.org/database/.

[40] SEDUE Ecoplan del Estado de Colima. Secretaría de Desarrollo Urbano y Ecología. 1980, 80 p.

[41] H. Lee, P. P. Hwang, H. C. Lin and F. L. Huang, *Fish Physiol. Biochem.*, 15, 513 (1996).

[42] T. Yada, T. Hirano and E. G. Grau, *Gen. Comp. Endocrinol.*, 93, 214 (1994).

[43] O. Watanabe, K. M., Burnett, B. L. Olla and R. I. Wicklund, *J. World Aquacult. Soc.*, 20, 223 (1989).

In: Effects and Expectations of Tilapia ... ISBN: 978-1-63463-307-9
Editors: M. A. Liñan Cabello et al.

Chapter 5

EVALUATION OF THE GROWTH OF TILAPIA IN PONDS AT MÉXICO CITY

J. L. Gómez-Márquez[1*], B. Peña-Mendoza[1], J. L. Guzmán-Santiago[1], E. Domínguez-de la Cruz[1] and A. G. Sánchez-Viazcán [1]

Laboratorio de Limnología, Facultad de Estudios Superiores Zaragoza, UNAM, México, D.F.

ABSTRACT

In countries that have been introduced the tilapias, they represent a considerable economic resource for its production in the different reservoirs and ponds. In Mexico, a production of 73,373 ton was obtained in 2009, ranking third place in aquaculture at the national level. In this chapter the growth of *Oreochromis niloticus* males sexually reversed was analyzed in concrete ponds in the FES Zaragoza, UNAM, under the environmental conditions of the Mexico City. The experiment was carried out from April to September 2010, in a concrete pond of 50 m^2, with three divisions. During this period, EL Pedregal balanced food was supplied daily at the rate of 6% of their biomass and total length and body weight was recorded monthly to get the type of growth. The following indicators were calculated: final body weight (FBW), final total length (FTL), percentage gain in weight and length (PGW or PGL), specific

[*] lgomez@unam.mx.

growth rate (SGR), feed conversion rate (FCR), Fulton's condition index (K) and survival. Water quality was monitored along the experiment. Based on the results, there were not statistically significant differences in size ($F=0.59$, $p>0.05$) and weight ($F=0.75$, $p> 0.05$) between the divisions of the pond. The initial mean total length used was 4 cm and ended with 21.3 cm. The initial average body weight of the fish was 1 g and they recorded 140 g the end of culture. The weight-length relationship for the fish was positive allometric ($b=3.099$, $p<0.05$). The percentage gain in body weight and total length averaged were 150% and 31.8% respectively. Specific growth rate in weight was of 3.0. Daily weight gain was of 0.52 g/day and 0.11 cm/day. The FCR had a mean value of 1.8:1 and condition factor (K) had an average of 1.45. Water quality indicate good oxygenation (> 5 mg/L), warm water (> 23°C), productive (<100 mg/L $CaCO_3$) and slightly basic (pH>8). Acceptable growth of fish and a yield of 3 ton/ha/6 months were obtained; therefore, the culture of tilapia under conditions of Mexico City is recommended.

INTRODUCTION

Tilapias of the family Cichlidae were introduced from Africa [1] and represent important social and economic benefits to rural communities [2]. The Nile tilapia, *Oreochromis niloticus* (Linnaeus) is one of the most important freshwater fish in world aquaculture [3, 4]. It is widely cultured in many tropical and subtropical countries of the world [5]. Rapid growth rates, high tolerance to adverse environmental conditions, efficient feed conversion, ease of spawning, resistance to disease and good consumer acceptance make it a suitable fish for culture [6, 7, and 8]. Farmed tilapia production decreased dramatically in recent years, decreasing from 92 981 ton in 1993 to 75,927 ton in 2011 [9]. The major concern for tilapia aquaculture is excessive reproduction and the resulting small size of the fish produced [10, 11, and 12]. Hence, the desirability of monosex male populations of tilapia is well established for increased production potential and low management requirements. Besides, sex-specific differences in growth are significant in Nile tilapia where males grow significantly faster, the metabolic energy is channeled towards growth larger and more uniform in size than females. They benefit from anabolism enhancing androgens. In females, there is a greater reallocation of metabolic energy towards reproduction [13, 14].

Intensive monoculture of the fish in concrete tanks is carried out in a few countries [15]. The problem of overpopulation in ponds caused by uncontrolled reproduction is a major constraint to the further development of the tilapia culture industry. One solution to this problem is the use of monosex male populations which can be produced by a number of means [15, 16, and 17]. One method commonly used in developed countries and increasingly gaining acceptance in developing countries is the culture of all-male fish produced by the oral administration of 17a-methyltestosterone (MT) during the period of gonadal differentiation of the fish [17, 18]. Then, the purpose of this chapter was to evaluate the growth of monosex stock of Nile tilapia in Mexico City, in an outdoor fertilized concrete pond with available natural and supplementary food.

MATERIAL AND METHODS

Experimental Fish and Feeding Trial

Nile Tilapia fry (*O. niloticus*) were obtained from Aquaculture Production Center Zacatepec, Morelos belongs to CONAPESCA, Mexico. Tilapia were transported in plastic bags filled with hatchery pond water and filled with air and transported to Experimental Aquaculture Unit, at the FES Zaragoza of the Universidad Nacional Autonoma of Mexico in Mexico City, where the experiment was realized. The batch of tilapia stocked in April has been over-wintered. One concrete pond with three divisions was used for stocking of the monosex tilapia, with a surface area of 50 m^2 and water depth of 0.75 m.

Initially, 350 fish fry were introduced in 1,000-L circular tank and maintained during 40 days before start the experiment. During this time they were fed with steelhead rainbow trout commercial feed (41.9±0.13% of protein and 12.08 ±0.12% of lipids). The acclimated fish (initial mean body weight 1.0±0.05 g) were placed into a concrete pond with three divisions. One hundred fingerlings (six fish / m^2) were randomly distributed in each division. Fish were fed with commercial steelhead rainbow trout feed for 180 days (April to September 2010) at the rate of 6% of their body weight for the first thirty days then gradually reduced to 3% for remaining days. The daily ration was given in two equal proportions at 10:00 and 17:00 h, respectively. At the beginning of the experiment and every 30 days, 40 fish fry was individually were measured for total length (TL) and standard length (SL) to the nearest 0.1 cm and weighed (body weight, BW) to the nearest 0.1 g. Water quality was monitored along the experiment and *in situ* measurements were taken for environmental variables such as: temperature and dissolved oxygen (DO) with a water oximeter using Hanna portable (model HI9146); transparency with a Secchi disk. Conductivity, pH and total dissolved solids were measured *in situ* using Hanna portable combo waterproof pH/EC/TDS/Temperature Tester, model HI 991300. Total alkalinity, total harness and ammonia-nitrogen were determined in the laboratory according to APHA, AWWA and APWA [19] and Arredondo and Ponce [20] procedures.

Growth Performance and Fed Efficiency

The following indicators were calculated: Initial Body Weight (IBW); Final Body Weight (FBW); Initial Total Length (ITL); Final Total Length (FTL); Daily Weight Gain (DWG); Daily Size Gain (DSG); Specific Growth Rate (SGR in %/day) = 100 x (Ln final weight (g) – Ln initial weight (g) /days of the experiment); Feed Conversion Rate (FCR) = feed intake (g)/weight gain (g); Feed Efficiency (FE) = weight gain (g)/Feed intake (g); Fulton's condition index (K) = (body weight/total length^b) x 100 and % survival according with Bushan and Banerjee [15], Nguyen and Little [21] and, Ergün et al. [22].

The total length (TL)-total body weight (BW) relationship was calculated by a power regression between these variables for each sex using the formula:

$$BW = a\,TL^{b}$$

Where BW is the body weight, TL the total length, *b* the growth exponent or length-weigh factor, and *a* is a constant. The values of a and b were estimated by means of a "linearized" form of that equation by taking (base 10) logarithms on both sides and estimating the values of log (a) and log (b) by means of a linear regression, using ordinary least-squares regression. Student's *t* test was used to accept (or reject) the hypothesis of isometric growth [23].

Statistical Analysis

Data were subjected to analyses of variance (one-way ANOVA) and multiple comparisons of means by Tukey's test [24], in order to determine significant differences on fish performance ($p<0.05$), using the statistical program SPSS Version 15.0 (SPSS, Michigan Avenue, Chicago, IL, USA).

RESULTS

Water Quality Parameters

The physical and chemical parameters of pond water take every third day during the experiment (Table 1) were maintained within the tolerance range of Nile tilapia. The water mean value temperature obtained in the experiment was of 23.5±0.1°C. Statistical difference at $P<0.05$ was noticed in the water temperature among the months. Secchi disc transparency with a mean value of 0.35±0.05 m was obtained during the study. Dissolved oxygen fluctuated between lowest of 5.2±0.25 mg/L obtained in August 2010 and the highest of 9.9±0.3 mg/L recorded in June. Statistical difference at $P<0.05$ was observed in the dissolved oxygen concentration among the months (Figure 1).

Table 1. Data of stocking and harvest parameters of Nile tilapia monosex male concrete ponds (mean values±s.e.) in period from March to November 2010

Parameters	Means of parameters	Range of parameters
Water temperature (°C)	23.5±0.1	22 – 28.8
Dissolve oxygen (mg/L)	10.3±0.25	6.3 – 16.6
pH	9.1±0.2	7.9 – 10.5
Secchi disk (m)	0.35±0.05	0.30 – 0.40
Total hardness (mg $CaCO_3$/L)	68±9	36 – 98
Conductivity (µS/cm)	890±25	749 - 1145
Total Dissolved Solids (mg/L)	725±14	520 - 952
Alkalinity (mg $CaCO_3$/L)	178±15	113 -225
Ammonia (mg/L)	0.28±0.15	0.12 – 0.55

The surface water pH fluctuated between slight to moderate alkalinity. No acidic pH was recorded; it was alkaline for most part of the study (Figure 1). The monthly mean variations in electrical conductivity and total dissolved solids (TDS) followed similar trend with increase to end of period. There were variation in conductivity and TDS during the study. Both electrical conductivity and TDS showed significant differences in their concentrations among the months. The range of the total hardness was of 35.8 mg $CaCO_3$/L to 98.2 mg $CaCO_3$/L. Total ammonia nitrogen concentration was lowest at 0.1±0.05 mg/L at the beginning of the study (May), it gradually increased until a maximum concentration of 0.5±0.05 mg/L in September.

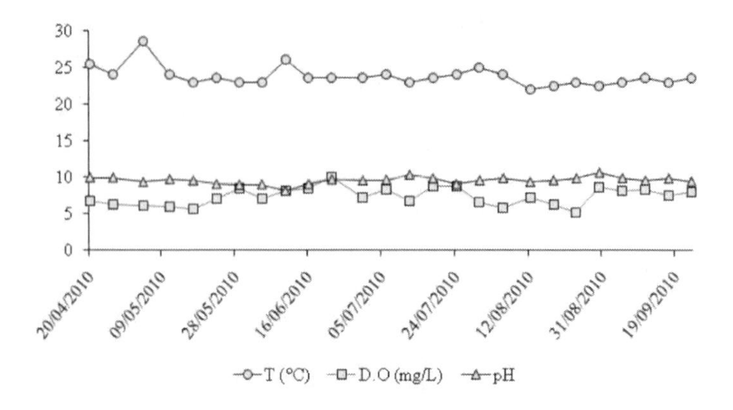

Figure 1. Values of physic-chemical parameters in concrete ponds during monosex tilapia culture.

Growth Performance

The tilapia fish maintained a trend of weight increase along the 180 days trial. However, higher increase was observed at 90 days with the commercial feed utilized. FBW, FE, DWG, FCR and K (Table 2) of fish were significantly affected and improved by protein level of the food and possibly by the water temperature range. It is important mentioned that the mean DWG of fish in all treatment ranged from 0.10 to 0.99 g/day (Table 2) and the FBW is considerate very good in this concrete pond in the Mexico City.

Table 2. Data of stocking and harvest parameters of Nile tilapia monosex male concrete ponds (mean values±s.e.) in period from March to November 2010

Growth parameters	Means of parameters	Range of parameters
Initial body weight (g)	1.0±0.05	0.8 - 1.2
Final body weight (g)	74.8±10.9	23.7 - 140.3
Initial total length (cm)	4.0±0.1	3.8 - 4.2
Final total length (cm)	15.5±1.3	11.0 - 21.3
SGR (%/day)	3.0±0.12	2.0 - 4.7
DWG (g/day)	0.52±0.15	0.10 - 0.99
DSG (cm/day)	0.11±0.05	0.10 – 0.12
FCR	1.8±0.03	0.7 - 2.3
FCI (K)	1.45±0.11	1.3 - 1.6
FE	1.03±0.09	0.81 – 1.21
Survival (%)	97.2	

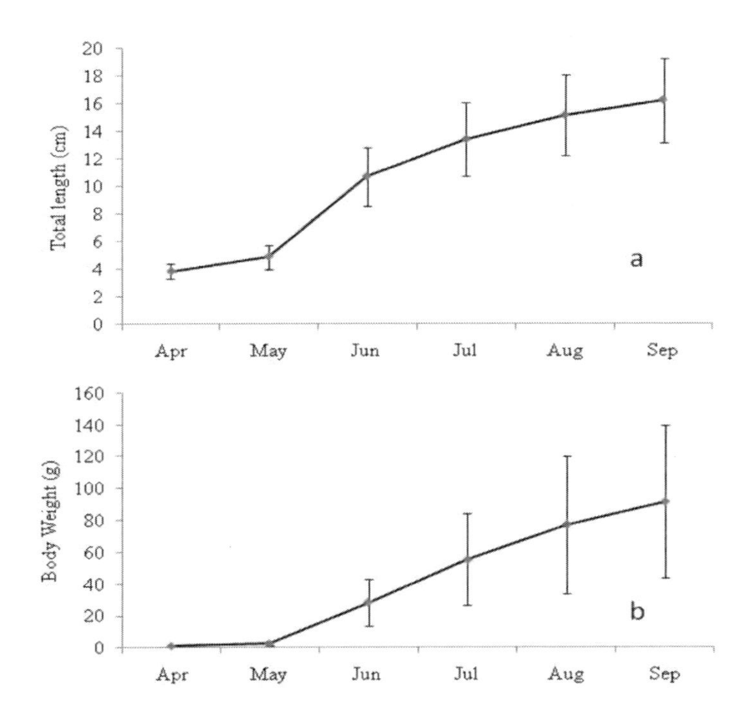

Figure 2. Variation of mean values of total length (a) and the body weight (b) of monosex tilapia (*O. niloticus*) culture.

The survival percentage in all the culture was around 97.2%, the lower mortality was registered in June, probably by the manipulation realized to the fish. Total sample size of *O. niloticus* was 292 individuals in 50 m^2 during 180 days. Total length and body weight ranged from 4.0 to 21.4 cm and 1.0 to 140 g respectively (Figure 2). The total length ($F=0.59$; $P>0.05$) and body weight ($F=0.75$; $P>0.05$) did not differs significantly between divisions.

Relationship between standard length and total length was statistically significant and the growth between parameters in not isometric (Figure 3):

$$SL=-0.021+0.791TL, (r^2=0.995, p<0.05)$$

Length-weight relationship was calculated for all individuals and this equation is given as follows:

$$BW = 0.013\ TL^{3.0.99}, (r^2 = 0.995, p<0.05)$$

Weight increases allometrically with size (Figure 4) since the b value was significantly different than 3 (t-value = 4.67; $p<0.05$).

Condition increased from April to September, maybe due to prevailing adequate environmental conditions. When the growth began in May, the condition of the males decreased; however in September the maximum condition values were achieved, as a consequence of weight increase due to the environmental conditions and food available.

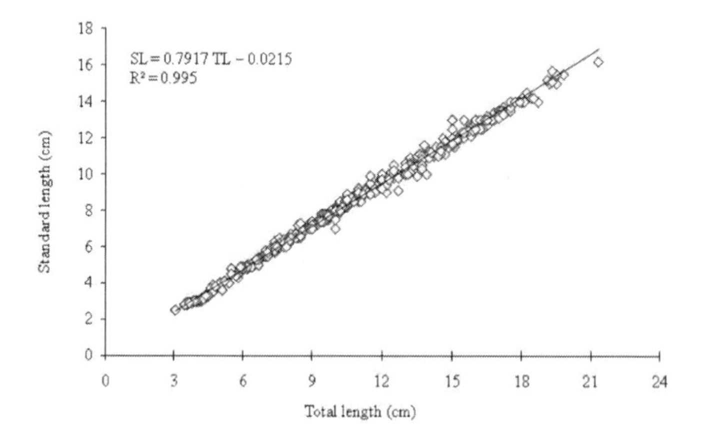

Figure 3. Standard length-total length relationship of monosex tilapia (*O. niloticus*) culture.

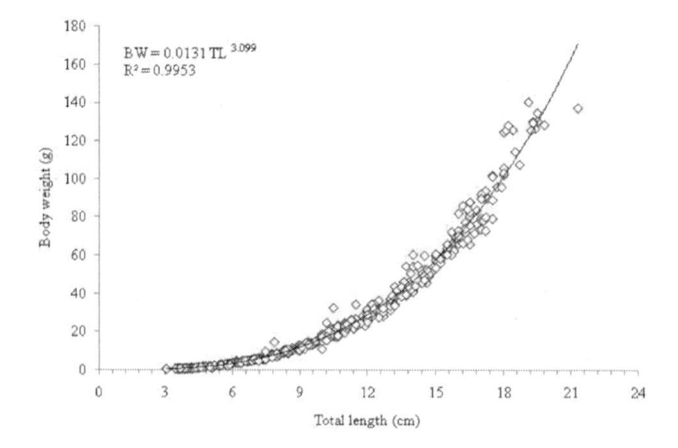

Figure 4. Length-weight relationship of monosex tilapia (*O. niloticus*) culture.

DISCUSSION

The potential role of tilapia culture in Mexico City was clearly indicated by the experiment. Fish of strains monosex reached a size of less of 200 g within a culture period of 6 months in ponds. Pandian and Sheela (1995) and Dagne et al. (2013) reported similar result, where all male tilapia showed superior growth rate over the females and mixed-sex which is in agreement with the results of the present chapter. They attributed this to the fact that energy is not utilized for reproduction and there exist no competition with younger fish in all male tilapia culture. However, has been registered that the tilapia of more than 200 g attracts premium prices in the market at national level and, it is more accepted by consumers.

The variation in weight observed in this chapter, might be explained by the establishment of hierarchies in feeding among the fish. Dominant individuals within a population may have consumed more food and have grown faster leaving less food for the submissive individuals, which in turn had slower growth and may have low biomass. Hierarchical interactions among fish was evidenced by the variation in size (few small and numerous large) of harvested fish following treatment. Similar results were reported by [18].

Monosex tilapia showed significantly higher weight, length, survival rate. The higher values of weight of the fish treated can be attributed to the anabolic effect of hormone to induce sex reversal in farmed tilapias [27]. In this respect, some studies reported that the anabolic effect of hormone showed an increase in individual growth of tilapia [28, 29, and 30]. Other studies reported that the higher mean weights could be attributed to the improvement of food conversion efficiency of sex-reversed fry of *Oreochromis niloticus* [26, 31].

Growth of individual fish was compared too with other studies. In the present chapter the growth rate was less to 1 g/day. Diana et al. [32, 33] observed individual growth of 1.2 g/day in hand-sexed male tilapia raised in ponds receiving high rate of fertilizers and 1.7–2.0 g/day in fertilized ponds with supplemental feeding. Siddiqui et al. [34] reported growth rates of 1.7–1.8 g/day in Nile tilapia reared in out-door tanks.

A further major factor affecting the growth rates of the fish was the relative exposure to lower temperatures occurring in the latter part of the growing season. Whereas fish grew during the period of optimal temperatures early in the season (May–July, 25–28°C), the growth rate were lowest when the ambient temperature began to decline toward the end of the culture (August–September; 24.5-22°C).

Monthly growth estimates were based on subsamples, while final fish size was determined from all fish in the pond. Thus, variation in tilapia production during the last month of the experiment was more likely due to biased estimates from previous months than to a different growth rate.

Vera Cruz and Mairb [18] cited that the better performance of hormone-treated fish compared to untreated fish during grow-out is in accordance due to a combination of factors: faster growth of males, improved food conversion efficiency and activation of other endogenous anabolic hormones enhancing growth.

Mair and Little [35] mentioned that using monosex seed can control the stunting problem caused by competition for food that otherwise occur between recruits and stocked fish. Moreover, has been showed that on average, monosex fish grew more than 10% faster than mixed-sex fish in both ponds and cages [29]. Faster growth of monosex tilapia has been related to the lack of energy expenditure in egg production and mouth brooding by females and lower energy expenditure on courtship by males (Macintosh and Little, 1995; cited in Dan and Little [29])

The results of the present chapter showed that the best weight gain was observed as 140.3 g, the specific growth rate (SGR) was recorded of 3.0 and the food conversion ratio (FCR) was 1.8. These values were similar to registered by Ahmed et al. [36] and Dagne et al. [26]. The commercial food presented highest values of FBW, SGR, FE and low FCR, showing that tilapia fish require a dietary protein level of 40 and 12% of lipids. This diet allows for a good growth and high survival rate (86.4 to 97.2%, respectively). These results are in agreement with the optimum P/E ratio for optimum growth of several species of teleost fish that range from 19 to 27 [37]. Comparatively, the requirements of tilapia fish are similar to those reported in other carnivorous fishes ranging between 32 to 50% of protein and 7 to 17% of lipids such as in the fingerlings of bay snook [38].

Wilson [39] point out that estimation of protein and lipids requirements of fish can be affected by factors such as rearing conditions, stage of growth and sources of protein, but more significantly by dietary energy content based on a quantitative determination.

It was observed in other experiments that fish treated at higher densities had similar total length to those treated at lower densities; thus it is apparent that the fish held at higher densities were in poorer condition [18]. A considerable advantage of the adequate or low stocking density was the significant reduction in mortality, and then high survival, as happened in this experiment.

Monthly mortality was estimated from known numbers of fish only at the beginning of the experiment. While mortality rate was assumed constant over time, losses may have occurred mainly over some small time interval. Mortality was generally small (<5%), so errors from this assumption should have been minor. Similar results were obtained by Diana et al. [32], but different to reported by Dange et al. [26].

Alkaline pH was also found to favour tilapia growth in the concrete pond together with the conductivity and total dissolved solids. The aquatic system also showed to be productive and will support the density of population of fishes. Then, the tilapia growth was strongly influenced by the physico-chemical factors which showed the water quality to be good according to APHA, AWWA and APWA [19] and Arredondo and Ponce [20]. Temperature, food abundance, nutrients were some of the factors that could limit fish growth in the pond, and maintenance of good water quality will be a great advantage for fish production.

Water temperature is one of the most important environmental factors affecting fish physiological responses of growth and feed utilization. Reports on Nile tilapia showed that the optimal temperature for growth and reproduction was between 22 to 32°C [40, 41]. For fish increased water temperature could increase both feed intake and growth, while excessively high temperatures result in decreased growth and feed efficiency with decreased feed intake [41, 42, and 43].

In conclusion, the present chapter revealed that the higher values of mean length, weight and survival rates were recorded in fish on sex reversal of *Oreochromis niloticus* in the environmental condition in Mexico City. It is possible that the higher survival at lower densities observed in this chapter was due to good water quality in the pond during the treatment period. Moreover, an alternative explanation for decreased growth could be deleterious pH levels caused by inorganic fertilizer addition, and the fish size could be affected by non optimal temperature during experimental period. Use of hormone to induce sex reversal in farmed tilapias has become a common practice, and it is a simple and reliable way to produce all-male tilapia stocks, which consistently grow to a larger uniform size than mixed sex or all-female stocks.

ACKNOWLEDGMENTS

We thank everyone who kindly assisted in several parts of the work for collecting, processing and providing measurements of *O. niloticus*. This work was supported by the Facultad de Estudios Superiores Zaragoza, U.N.A.M.

REFERENCES

[1] Gómez-Márquez, J.L., B. Peña-Mendoza, I.H. Salgado-Ugarte and M. Guzmán-Arroyo. *Mexico. Rev. Biol. Trop.* 51, 221 (2003).
[2] Jiménez-Badillo, L. *Rev. Biol. Trop.* 54, 577 (2006).
[3] Admassu, D. *Hidrobiología.* 337, 77 (1996).
[4] Coward, K. y N.R. Bromage. *J. Fish Biol.* 53, 285 (1998).
[5] Lin, Y.H., S.M. Lin and S.Y. Shiau. *Aquaculture.* 284, 207 (2008).
[6] Wohlfarth G.W. and G. Hulata. Applied of Tilapias. *ICLARM Studies and Reviews* 6. (1983).
[7] Yi, Y, C.K. Lin and J.S. Diana. *Aquaculture.* 146, 205 (1996).
[8] El-Saidy, D.M.S.D. and M.M.A. Gaber. *Aquaculture Res.* 36, 163 (2005).
[9] CONAPESCA. Anuario Estadístico de Acuacultura y Pesca 2011. Comisión Nacional de Acuacultura y Pesca. SAGARPA, México. (2012).
[10] Babiker, M. and H. Ibrahim. *J. Fish Biol.* 14, 437 (1979).
[11] de Graaf, G.J., F. Galemoni and E.A. Huisman. *Aquaculture Res.* 30, 25 (1999).
[12] Lèveque, C. *Environ. Biol. Fishes.* 64, 461 (2002).
[13] Angienda, P.O., B.O. Aketch and E.N. Waindi. *International Journal of Biological and Life Sciences.* 6(1), 38 (2010).
[14] El-Greisy, Z.A. and A.E. El-Gamal. *The Egyptian J. of Aquatic Res.* 38(1), 59 (2012).
[15] Bushan, S.C. and S. Banerjee. *World J. of Fish and Mar. Sci.* 1(3), 212 (2009).
[16] Buddle, C.R. *Aquaculture.* 40, 233 (1984).
[17] Baltazar, G.P.M. *INFOPESCA INTERNACIONAL.* 40, 21 (2009).
[18] Vera Cruz, E.M. and G.C. Mairb. *Aquaculture.* 122, 237 (1994).
[19] APHA, AWWA y WPCF. Standard Methods for Examination of Water and Sewage and Wastewater. 18ª ed. EE.UU. (1992).

[20] Arredondo, F.J.L. and J.T. P Ponce. Calidad del agua en acuicultura: Conceptos y aplicaciones. AGT Editor.S.A. (1998).

[21] Nguyen, C.D. and D.C. Little. *Aquaculture*. 184, 221 (2000).

[22] Ergün, S., D. Guroy, H. Tekesoglu, B. Guroy, I. Celic, A.A. Tekinay and M. Bulut. *Turkish J. Fish. Aquat. Sci.* 10, 27 (2010).

[23] Pauly, D. Fish population dynamics in tropical waters: A manual for use with programmable calculators. ICLARM. *Studies and Reviews 8, Manila,* Philippines. (1984).

[24] Steel, D.G.R. and J.H. Torrie. Principles and Procedures of Statistics. A biometrical approach. *McGraw-Hill,* Nueva York, USA. (1981).

[25] Pandian T. J., Sheela S. G. *Aquaculture*. 138, 1 (1995).

[26] Dagne, A., F. Degefu and A. Lakew. *Ethiopian Inter. J. of Aquaculture.* 3(7), 30 (2013).

[27] Jo, J.Y., R.O. Smitherman and D. Tave. *J. of Aquaculture.* 8(2), 77 (1995).

[28] Mair, G.C., J.S. Abucay, J.A. Beardmore, D.O.F. Skibinski. *Aquaculture.* 137, 313 (1995).

[29] Dan, N.C. and D.C. Little. *Aquaculture.* 10(2), 32 (2000).

[30] Little, D.C., R.C. Bhujel and T.A. Pham. *Aquaculture.* 221, 265 (2003).

[31] Chakraborty, S.B. and S. Banerjee. *Inter. J. of Biol.* 2(1), 44 (2010).

[32] Diana, J.S., C.K. Lin and P.J. Schneeberger. *Aquaculture.* 92, 323 (1991).

[33] Diana, J.S., C.K. Lin and K. Jaiyen. *Journal of World Aquaculture Society.* 25(4), 497 (1994).

[34] Siddiqui, A.Q., M.S. Howlader and A.B. Adam. *Aquaculture and Fisheries Management.* 20, 49 (1989).

[35]] Mair, G.C. and D.C. Little. NAGA, The ICLARM. *Quarterly* 4(2), 8 (1991).

[36] Ahmed, G.U., N. Sultana, M. Shamsuddin and M.B. Hossain. *Pak. J. Biol. Sci.* 16(23), 1781 (2013).

[37] Nutritional Research Council (NRC). Nutrient requirements of fish. *National Academy of Press,* Washington, D.C. (1993).

[38] Arredondo-Figueroa, J.L., 1J.J. Matsumoto-Soulé, J.T. Ponce-Palafox, K. Shirai-Matsumoto and J.L. Gómez-Márquez. *International Journal of Animal and Veterinary Advances.* 4(3), 204 (2012).

[39]] Wilson, R.P. Amino Acids and Proteins. 1989. Halver, J.E. (Ed.), *Fish Nutrition, Academic Press,* New York, pp 111-151.

[40] Morales, D. A. La Tilapia en México. Biología, Cultivo y Pesquerías. AG Editor, S.A. (1991).

[41] El-Sayed, A.F.M. and M. Kawanna. *Aquaculture Res.* 39, 670 (2008).
[42] Handeland, S.O., A.K. Imsland and S.O. Stefansson. *Aquaculture.* 283, 36 (2008).
[43] Xie, S., K. Zheng, J. Chen, Z. Zhang, X Zhu and Y Yang. *Aquaculture Nutrition.* 17: e583 (2011).

In: Effects and Expectations of Tilapia … ISBN: 978-1-63463-307-9
Editors: M. A. Liñan Cabello et al.

Chapter 6

PROJECTIONS OF THE EXPECTATION *PANGASIUS HYPOPHTHALMUS* CULTURE ON *OREOCHROMIS SPP* IN TROPICAL COUNTRIES

***Laura A. Flores-Ramírez*[1,*]**
***and Marco A. Liñán Cabello*[2]**
[1]Centro Regional de Investigación e Innovación Pesquera y Acuícola, CRIIPA-Manzanillo, Instituto Nacional de Pesca, Manzanillo, Colima, México
[2]Biotecnología, FACIMAR, Universidad de Colima, Manzanillo, Colima, México

ABSTRACT

Tilapia and *Pangasius* are today the two dominant farmed whitefish species in the world market. Rapid increases in production tilapia can be attributed to improvements in aquaculture technology and infrastructure in several nations in the Americas that are major producers, and to greater numbers of trained biologists. *Pangasius* (basa and tra) culture is the powerhouse in the freshwater aquaculture sector and is probably the most successful national aquaculture industry to date in terms of growth rate of exports and speed of market diversification. The advantages in the culture

* Corresponding author: Laura A. Flores-Ramirez, Email address: lauflores64@gmail.com.

of these species represent a potential opportunity for tropical countries to supply this growing demand not only species but committed to diversification of aquaculture. The challenge comes to appropriate and apply new techniques and improving those already made under performance-based set of standards for the industry to come as a sustainable source of seafood. The market integration and thus competition will play a key role for the expectations of projection in tilapia and *Pangasius* species.

INTRODUCTION

World aquaculture production has emerged as one of the fastest growing food sectors and this output now makes an increasingly significant contribution and this trend is projected to continue.

In these terms, global aquaculture has been dominated by higher valued, marine fish species, namely, salmon, trout, sea bass, sea bream and mussels. However, freshwaters were the source for 60 per cent of the world aquaculture production in 2008, dominated by various species of carp, although tilapia and later *Pangasius* catfish have become more significant [1]. Asia and China has the greatest freshwater aquaculture production in relation to land area, although some European and African countries are also significant. The Americas in particular are notable for relatively low freshwater aquaculture production per unit area [1].

At this respect, tilapia and *Pangasius* are today the two dominant farmed whitefish species in the world market. Pangasius has experienced much higher growth rates than tilapia during recent years. From to 2002 to 2008, average annual production growth was enclose to 50% for *Pangasius*, but only 8 to 9 % for tilapia. Raised in more than 100 countries, tilapia is the most diversified market base. On the other hand, the Vietnamese *Pangasius* industry is probably the most successful national aquaculture industry to date in terms of growth rate of exports and speed of market diversification [2].

Moreover, in the last two decades with significant price increases from the mid-1990s while the highly segmented whitefish market has, perforce, been expanding to include a series of new species, e.g. tilapia and *Pangasius*, beyond the traditional species. The last will increase their demand and most of these increases will occur in tropical regions. Here, tropical countries have a potential opportunity to diversifying its aquaculture.

The challenge comes to the development or transfer of appropriate technology and a favorable business environment under the guidelines of better management practices that generate safe products, reducing negative impacts environmental and adequate quality controls for export. The market integration and thus competition will play a key role for the expectations of projection in this important species.

Global Tilapia Production

Tilapia rank high in global aquaculture production. Carps are the only category of fish species with greater production than tilapias (Figure 1). Tilapias are a hardy species produced by several culture methods under a wide range of environmental conditions. They are tropical and subtropical species, but they have been cultured at temperate sites by using geothermal water, greenhouses, or other means of providing warm water during winter. Tilapias are produced in many countries, but most production occurs in tropical and subtropical regions in developing countries. Contrary to some aquaculture species, tilapias are important in local and export markets as well as food fish by rural farmers [3].

While significant worldwide distribution of tilapias, primarily *Oreochromis mossambicus*, occurred during the 1940s and 1950s, distribution of the more desirable Nile tilapia occurred during the 1960s up to the 1980s [4]. Nile tilapia was introduced to China, which leads the world in tilapia production and consistently produced more than half of the global production in every year from 1992 to 2008 [3]. The development of hormonal sex-reversal techniques in the 1970s represented a major breakthrough that allowed male monosex populations to be raised to uniform, marketable sizes. In addition, research on nutrition and culture systems, along with market development and processing advances, led to rapid expansion of the industry since the mid 1980s [5]. Several species of tilapia are cultured commercially, but Nile tilapia is the predominant cultured species worldwide [6].

Tilapia is the production of an estimated 100 countries. China is the largest producer with its production reaching almost one million metric ton in 2005 and also the largest exporter of tilapia products, having overtaken. Total tilapia production in the Americas was approximately 204,267 tonnes in 1998. Tilapia production increased 13%/year from 1984–1995. Production of *O. niloticus* and red tilapia have experienced the highest growth rates throughout the Americas due to high prices in markets and the high demand for large fillets. Mexico, Brazil and Cuba are the three largest producers [4]. In 2004 the total world production of tilapias and other cichlids reached 2.6 million tonnes and continued to rise to up to 3.6 million tonnes in 2008. Production of tilapias has a wide distribution, and 72 percent are raised in Asia (particularly in China and Southeast Asia), 19 percent in Africa, and 9 percent in America [3].

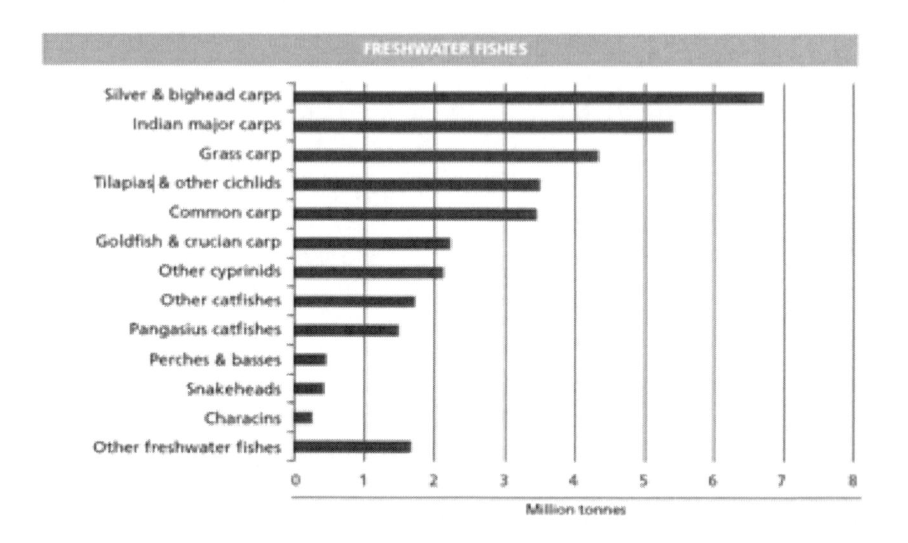

Figure 1. Production of major species or species group from aquaculture in 2010 [3].

There is a strong demand for tilapia in the United States. U.S. tilapia consumption is around 360,000 tons. Demand for frozen fillet imports, which mainly come from China and other Asian countries, continues to grow strongly, with an increase of 1500 units since 2006. Whole frozen tilapia remains stable from 2006 to 2007. Fresh tilapia fillets, which mainly come from Latin American countries, increased 20% [7].

The countries of the Americas are relatively small markets and producers compared to China and other Asian countries. However, the US is a rapidly growing market that has encouraged tilapia farms to develop throughout the Americas. Growth of domestic markets in South and Central America has further supported demand. Mexico and Brazil in particular have strong domestic markets. These two countries with large populations and enormous aquatic resources will be the two major players in the Western Hemisphere tilapia industry. Tilapia markets throughout the Americas will diversify as more value-added products are offered [4].

At this respect, the demand for certified tilapia is growing in European markets. Malaysian tilapia producer Trapia Malaysia has been certified by the Aquaculture Stewardship Council. The company uses GenoMar Supreme

Tilapia fingerlings from its onsite hatchery next to Lake Temenggor in the north peninsula of Malaysia. The certified fish are sold as frozen fillets and loins to North America, Europe and Asia, and as live and fresh in the local market. Certified tilapia has been available in the market since August 2012 [8].

PANGASIUS BIOLOGY

Pangasius belongs to the family of catfish in the Mekong Delta, composed of the Mekong River and its estuaries in Southeast Asia. There are two species of catfishes being cultured in Viet Nam, namely *Pangasius hypophthalmus* (tra) and *Pangasius bocourti* (basa). *P. bocourti* used to be the main culture species in the early days, then culture of *P. hypophthalmus* took over since the beginning of 2000, becoming the main cultured species and accounting for more than 95% of total aquaculture catfishes. At present the production of *P. bocourti* remains only 3% of the total catfishes production [9].

Sexual maturation in *P. hypophthalmus* takes more than three years, at least in captivity [10]. In nature it may well be the same. It is not clear at what size the species reaches sexual maturity [11].

It can reach 54 cm in length and a minimum weight of 3-4 kg [12]. Females are larger than males in their natural environment and egg production is seasonal and occurring during the warmer months [9]. *P. hypophthalmus* is a prolific spawner. Egg production per kilo of fresh weight increases dramatically with increasing size until a weight of around 10 kg has been reached [11]. Apparently some fish are capable of spawning twice in a year [10]. The eggs are sticky and are apparently deposited on roots of trees that become flooded early in the wet season [12]. The eggs hatch within 24 hours. It is not clear where the larvae go after hatching. After two or three days, they start feeding and by the time the larvae have developed into fry and fingerlings, they are pelagic and capable of independent movement [11].

As the fish has an accessory respiratory organ and can also breathe with air bladder and skin, they can bear in the water lacking dissolved oxygen. This enables the fish to tolerate poor water quality, including high organic matter or low dissolved oxygen levels, and they can therefore be stocked at high densities [11]. The oxygen consumption and activity level of *Pangasius* is three times lower than of a silver barb [13]. The raising period varies from 6 to 8 months to reach the weight of 0.8 to 1.3 kg/fish. *Pangasius* fish can live in hot temperature up to 39 °C but easily died at temperature below 15 °C [13].

Catfish are omnivorous and will accept trash fish, pellets, home-made feed formulated from agro- and fishery by-products, water plants and even animal and human wastes when cultured in ponds and cages [11]. Its muscle is white with high nutritious value, a little of tasted lipid content, without horizontal bones and without smells of sediments and seaweed [7]. *Pangasius* is available in the same forms available for most fish. The most popular form is boneless, skinless fillets or portions in different sizes and shapes cut from fillets. Fillets can range in size up to 6-8 ounces. Most products are shipped to the U.S. frozen and are available as a frozen item or thawed and sold as a previously frozen refrigerated product.

The only factor that has limited the success of this fish as aquaculture product is not reproduced naturally in captivity shortly. *Pangasius* is played only in certain areas of the Mekong River and requires specific environmental factors to achieve. In the mid-nineties, French biologists, with the help of the government of Vietnam, managed to apply techniques to reproduce induced fish spawning hormonal nursery conditions, eliminating the dependence on wild harvest. This made the industry grow rapidly breeding *Pangasius* in Vietnam, in addition to achieving the goal of developing an export market [9].

PANGASIUS INDUSTRY

Main producer countries of *P. hypophthalmus* are Vietnam, India and Indonesia, and a lesser extent Thailand, Malaysia, Cambodia, Laos, China, Myanmar and Bangladesh (Figure 2). Currently Puerto Rico Puerto Rico is the only country in the Western Hemisphere where it is stabilized production as *Pangasius* fish consumption. Many other tropical countries in Latin America are interested in this potential but do not have the broodstock neither the necessary technology [9].

The *Pangasius* farming in 10 years increased very fast to become one of the species "top" of world aquaculture, reaching the same level of salmon and tilapia. Projections indicate that in a few years will be the first [3]. Vietnam being responsible for 90% of world exports (Table 1)

Principal markets for *Pangasius* have been established in Europe, the United States and Russia, although the product is exported to more than 100 countries worldwide [13]. Most recently, the fastest-growing markets have been in Central and South America. Mexico is now the fifth-largest importer of *Pangasius* in the world, while other Latin American countries such as Colombia, Brazil and Costa Rica have shown notable increases in recent years.

Ironically, these countries are also recognized as large-scale producers of tilapia for domestic consumption and export [9].

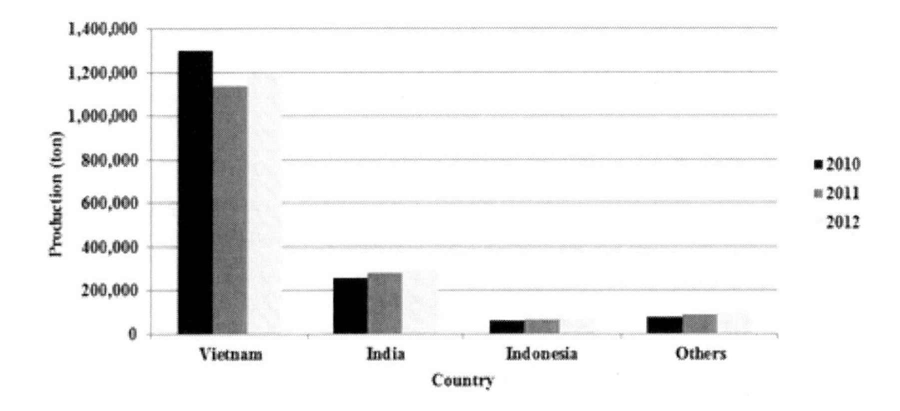

Figure 2. Global production of *Pangasius* sp.

Table 1. Main exporters of Vietnam in 2012

Exporting	$USD
Vinh Hoan Corp	154,988,396
Hung Vuong Corp	111,900,532
Agifish	91,857,355
Anvifish Co	82,787,234
I.D.I. Corp	58,225,736
Navico	58,051,547
NTSF Seafoods	50,176,911
HungCa Co. Ltd.	43,495,411
Dathaco	43, 211,162
CL-Fish Corp	42,734,081

The large-scale production in Vietnam as well as the production of neighboring China, has resulted in the marketing of *Pangasius* fillets at low prices which contributed to its rapid growth and acceptance in global markets [9]. *Pangasius* export in 2010 reached 659,000 MT, worth about US $1.427 billion, up 7.4 percent in volume and 5.2 percent in value compared to 2009 [13]. Also, the *Pangasius* industry has used decreasing prices as a means to access new markets, recent increases in production costs associated to higher energy and food costs have destabilized the Vietnamese industry while markets adjust to higher prices. This will likely create opportunities for the development of *Pangasius* in other countries, including the tropical regions of the Western Hemisphere [2, 9].

Pangasius industry provides significant quantities of food for human consumption, as fish is a key source of protein for many people. The industry also creates millions of jobs on and off the farm. As with any rapidly growing industry, the growth in aquaculture production often raises concerns about negative social and environmental impacts, such as water pollution, the spread of diseases and unfair labor practices at farms. And as in any industry, there are some businesses addressing these issues well and some that are not doing so at all or are doing so poorly. It is important that we face the challenge of promoting practices that contribute to resolving these issues, while reducing those that have a negative impact [14].

One solution to this challenge is creating voluntary standards for aquaculture production, as well as a process for certifying producers who adopt the standards. Such certified products can help reassure retailers, restaurants, food service companies and other seafood buyers that the aquaculture products they purchase were produced responsibly. Through a multi-stakeholder process called the *Pangasius* Aquaculture Dialogue (PAD), science-based standards for the *Pangasius* aquaculture industry have been created. The standards, when adopted, will help minimize the key negative environmental and social issues associated with *Pangasius* farming [14].

Particularly for tilapia, diverse production environments, methods, competencies and regulations give rise to differences in quality and production costs. This causes producers to target specific markets with different requirements, and contributes to segmentation. Since tilapia production is fairly small in many developing countries, it is often hard to build industrial clusters that can satisfy the high requirements of governments and buyers in developed markets and be competitive in developed markets.

TILAPIA AND *PANGASIUS* EXPECTATIONS

If *Pangasius* farming in tropical countries is discarded, the current industry based on tilapia could be exposed to the risk of increased competition, lack of product diversification. Thus, in real terms, it appears that the *Pangasius* has a role to play in strengthening and diversification of the aquaculture industry, and should be taken into account that opportunity to assess all the potential (Table 2).

Table 2. Comparison of advantages and disadvantages in tilapia and *Pangasius* farming

Tilapia	*Pangasius*
Exotic species	Exotic species
The use of hormone 17α-methyltestosterone for tilapia fries is the most simple and reliable way to produce all male	The application of methods for hormone induced spawning has been the principal factor leading to the rapid development of *Pangasius* aquaculture
Tilapia reproductive strategy include precocious maturation, frequent spawning, extended parental care, and non dependence by fry on specialized food resources.	*Pangasius* are highly fecund and a single female can produce more than 50,000 eggs per kg per spawn. *Pangasius* grow to large size and require 2 – 3 years to reach sexual maturity
Survival of Tilapia fingerlings is normally high assuming proper management and the avoidance of catastrophic bacterial or parasitic infections.	Low survival rates in juvenile, cannibalism, improper handling and insect attack.
It is omnivorous, it can be predatory. Build nests and territorial behavior is affecting the habitat.	It is omnivorous, benthic detritivore, not predatory or territorial or build nests. Not prove parental care
It can survive exposure to 0.6 mg/L of dissolved oxygen (DO) for short periods. Chronic exposure to 3 mg/L or less DO reduce growth.	This species can use a modified swim bladder to get oxygen directly from the atmosphere
Depends on the DO in water	This fish is not dependent on DO in water
Mechanical aeration may increase production costs significantly	Mechanical aeration is optional
Can not tolerate temperatures below 20 °C for extended periods	Can not tolerate temperatures below 20 °C for extended periods
Tilapia yields from intensively managed ponds range from 8–15 mt/ha	Intensive production of *Pangasius* in ponds can produce yields of 250 – 300 mt/ha

The increasing imports of *Pangasius* to tropical aquaculture producing nations of the Western Hemisphere present an immediate challenge to an aquaculture industry that would likely benefit from diversification in order to sustain development. However this introduced product could negatively impact future aquaculture development, established producers, domestic sales and revenues generated from exports. The challenge to tropical producing nations in the west is how best to respond to the growing *Pangasius* imports [9]. Three alternatives can be considered: Allow *Pangasius* imports to continue and wait to see at what point the markets stabilize. Pursue protectionist policies through the application of tariffs or changes in sanitary regulations, or by creating a negative image of the *Pangasius* product. Compete by considering the introduction of *P. hypophthalmus* as a new aquaculture species [9].

By the simple fact that the *Pangasius* is an aquaculture species superior in terms of production and distribution in the market, it can be considered as a feasible third option: to introduction of *Pangasius* in suitable for production in the Western Hemisphere tropical areas. The introduction of a new species for aquaculture always raises concerns related to potential environmental impacts, it is useful to compare the species with tilapia, which were widely introduced in the region over the last 40 years. Tilapia exhibit many positive traits as an aquaculture species but are also recognized as a highly invasive species. Despite these negative attributes, tilapia introductions have generally been considered positive when socioeconomic factors as well as environmental issues are assessed.

Pangasius have not been introduced for aquaculture outside tropical regions of Asia, although they are available as an ornamental species for the aquarium trade in many countries. Even in their native range, reproduction for aquaculture purposes is dependent on hormone-induced spawning. Currently, *Pangasius* are cultured at Caribe Fisheries Inc. in Puerto Rico and have recently been introduced to the Dominican Republic, Haiti and Jamaica.

Some countries are seriously considering their potential, which could benefit the diversification of aquaculture. The fact that there still exist *Pangasius* culture in the Western Hemisphere seems to be related, first, to initiatives from the private sector, which can not or should not start growing without the approval of their respective governments, and on the other, this potential is limited by three factors: the introduction of the species, transfer of technology and market development [9].

Direct competition between tilapia and *Pangasius* has so far been limited, because both species are helping to fill the void generated by declining harvests of traditional marine fisheries. As the aquaculture industry continues to grow and mature, it is likely that increased competition between these two species occupy similar market share [2].

At this respect, two markets are integrated if buyers and sellers can exploit arbitrage opportunities caused by changes in the price difference between the markets. An indication of integrated markets is that their market prices fluctuate together, even though the price levels may be different due to transportation costs and other factors. The degree of market integration between tilapia and *Pangasius* is also limited. However, this does not mean there are no competition or arbitrage opportunities across products or countries. Even if tilapia and *Pangasius* products are not perfect substitutes, there are threshold levels for price differences at which the substitution effect kicks in. This process is likely to be helped by the fact that the whitefish market seems increasingly globalized and integrated, and there is a stronger emphasis on the taste of the whitefish fillets rather than the species from which they are obtained [2].

In this sense, tilapia producers will face this situation, determine the importance of *Pangasius* in the future of aquaculture in tropical countries and together with their governments to develop strategies to adapt to the trends of the industry and market. However, it is not an easy task, since tilapia within certain limitations still exist. An example of this is made Latin countries, which have felt the impact of the Asian tilapia, which, though lower quality, has managed to displace its low cost tilapia produced in Latin America. The idiosincransia and the frenetic importing species that offer the most extraordinary panaceas are other factors to consider.

The farming of this species is a very high risk activity for household-scale production units because profit margins are extremely tight. It is likely in the medium- to long-term that household-scale grow-out producers will go out of business as the industry consolidates. Associated with increasingly strict market requirements for tracking, tracing and product certification, striped catfish farming will probably move toward large-scale vertically integrated operations.

Regulations from national and local governments and assistance from multinational organizations or buyers in developed countries may be needed to overcome trade barriers for several tilapia-producing countries. Yet, the ability to satisfy requirements of governments and buyers in developed countries will play a key role for future *Pangasius* and tilapia market integration.

CONCLUSION

The projections anticipate an increased in demand of fish protein where tilapia and *Pangasius* will play a key role. The last represent a potential opportunity for tropical countries to supply this growing demand and to diversification its aquaculture. The future development of tilapia and *Pangasius* aquaculture in the tropical regions will depend on the ability of transfer appropriate technology and favorable business environment production systems to produce more fish with less water, less food, and less time to lower production costs and reduce pollutants to the environment. At this respect, the development of appropriate standards can, however, be challenging. Within aquaculture, there are now many initiatives, perhaps most significantly GLOBALGAP, which is private sector-based business-to-business certification focusing on food safety, animal welfare, environmental protection and social risk assessment standards.

REFERENCES

[1] J. Bostock, B. McAndrew, R. Richards, K. Jauncey, T. Telfer, K. Lorenzen, D. Little, L. Ross, N. Handisyde, I. Gatward and R. Corner, *Phil. Trans. R. Soc. B*. 365, 2897 (2010).

[2] R. Tveterås, *G. Aquaculture Advocate*. 12, 37 (2009).

[3] FAO, The state of world fisheries and aquaculture, Rome, Italy (2012).

[4] K. Fitzsimmons, *Future trends of tilapia aquaculture in the Americas*, The World Aquaculture Society, Baton Rouge, Louisiana, United States (2000).

[5] W. O. Watanabe, T. M. Losordo, K. Fitzsimmons and F. Hanley, *Rev. Fish. Sci*. 10, 465 (2002).

[6] C. C. Kohler. Preprint, 2004 at http://www.ncrac.org/files/technical_bulletins/TilapiaWhitePaper11404.pdf

[7] J. Ogonoswki. Preprint, 2008 at http://pdf.usaid.gov/pdf_docs/Pnadn735.pdf

[8] Globefish, http://www.globefish.org/tilapia-june-2013.html

[9] M. V. McGee. Preprint, 2010 at http://www.thefishsite.com/articles/1013/pangasius-for-western-aquaculture

[10] P. Cacot, *Etude du cycle sexuel et maitrise de la réproduction de Pangasius bocourti (Sauvage, 1880) et Pangasianodon hypophthalmus*

(Sauvage, 1878) dans le delta du Mekong au Viet Nam, Ph. D. Thesis. Institute National Agronomique, Paris-Grignon, France (1999).

[11] N. Van Zalinge, L. Sopha, N. P. Bun, H. Kong and J. V. Jørgensen, *Status of the Mekong Pangasianodon hypophthalmus resources, with special reference to the stock shared between Cambodia and Viet Nam*, MRC Technical Paper No. 1, Mekong River Commission, Phnom Penh (2002).

[12] S. T. Touch, Life cycle of Pangasianodon hypophthalmus and the impact of catch and culture, Paper presented at the Catfish Asia Conference, Bogor, Indonesia (2000).

[13] Vietnam Association of Seafood Exporters and Producers (VASEP), Pangasius 26 Q & A, Agricultural Publishing House, Ha Noi, Vietnam (2012).

[14] Pangasius Aquaculture Dialogue Standards (PADS), Mohammad Mahfujul Haque, Bangladesh Agricultural University (2010).

About the Editors

Editor:
Dr. Marco Agustin Liñan Cabello
Doctor in Sciences, Marine Biotechnology,
Center for Scientific Research and Higher Education of Ensenada,
Professor-Researcher, Faculty of Marine Science, University of Colima,
Km 1 9.5 Carretera Manzanillo-Barra DeNavidad
Manzanillo, Colima, Mexico CP 28040
Tel: 3311200
e-mail: linanm@ucol.mx, Iinanmarco@hotmail.com

Co-Editor:
Dr. Laura A. Flores-Ramírez
Centro Regional de lnvestigacíon e lnnovacíon Pesqueray Acufcola,
CRIIPA -Manzanillo, lnstituto Nacional de Pesca,
Biotecnologfa, FAC IMAR, Universidad de Colima,
Manzanillo, Colima, Mexico
e-mail: Iauflores64@gmail.com

Index

A

B

C

D

E

F

G

H

I

J

L

O

P

Q

R

S

T

U

V

W

Y

Z